The Science of
SOAP FILMS AND SOAP BUBBLES

CYRIL ISENBERG
University of Kent at Canterbury

TIETO LTD.
5 Elton Road, Clevedon, Avon, England.

BL British Library Cataloguing in Publication Data

Isenberg, Cyril
  The science of soap films and soap bubbles.
  1. Soap-bubbles   2. Soap
  I. Title
  532'.6   QC183

  ISBN 0-905028-02-3

Tieto Ltd.,
5, Elton Road,
Clevedon, Avon BS21 7RA
England.

Printed in England by Woodspring Press Ltd., Church Street,
Bridgwater, Somerset.
ISBN O-905028-02-3

Dedicated to
Charles Vernon Boys

FOREWORD

There are few physical objects as beautiful or as fascinating as a soap bubble —fire is perhaps the only common natural phenomenon which can hold our interest in the same way. But, unlike fire, soap bubbles exhibit a perfection of geometrical form and a simplicity which appeals instantly to the mathematical mind.

The attraction of the subject to children and to scholars alike has been evident for over a century. Sir James Dewar, who was Resident Professor at the Royal Institution for forty-six years, gave the Christmas Lectures to young people in 1878–9 on "A Soap Bubble" and gave his last Evening Discourse in 1923, the year of his death, on "Soap Films as Detectors". C. V. Boys, who published his popular little classic on the subject in 1890, gave the Royal Institution Christmas Lectures in 1899 and A. S. C. Lawrence, who was assistant to Dewar, wrote his monograph on soap films in 1929 and was still studying them enthusiastically when he and I were colleagues in Sheffield during the 1960's. Clearly, it is easy to form a lifelong attachment to the subject.

The shapes and colours of soap films illustrate important principles and results in physical science and mathematics. For example, subjects concerned with the white light interference patterns of draining films, the shapes of minimum area surfaces contained by frameworks and the vibrational oscillations of soap film membranes are discussed here at levels that should be appreciated by the more advanced students of science. The book bridges the gap between the popular account of C. V. Boys and research texts, so it will appeal particularly to the more able sixth formers, as well as undergraduates and graduates. Soap films are closely related to biological membranes, which have assumed very great importance in recent years and this book helps to unify the sciences by discussing chemical, physical, mathematical and biological facets of the subject.

Those who have seen the colour plates in this book and particularly those of us who have seen Dr. Isenberg's skilful and artistic manipulation of soap films, during one of his lectures, will now be tempted to return to our early adventures in the blowing of bubbles and to repeat the experiments in a more sophisticated and enlightened way.

GEORGE PORTER
The Royal Institution London

PREFACE

Everyone has been fascinated, from an early age, by soap bubbles and soap films. This has been no less true of the scientific community. Biologists, chemists, mathematicians and physicists have all interested themselves in the properties of bubbles and films.

Professor Charles Vernon Boys is perhaps the best known popularizer of these properties. He gave numerous lectures and demonstrations at the end of the last century and the beginning of this century. His book *Soap Bubbles and the forces which mould them* was based on three lectures given to young people at the London Institution during the years 1889 and 1890. Since those days it has been studied by many generations of enquiring school children all over the world.

His book was written primarily for students aged ten to fourteen years. Consequently it does not discuss in depth such subjects as molecular structure, interference phenomena, and mathematical properties. This book is intended for the older student, or adult, with an undergraduate background in science, or at least a sixth form education, who would like to gain a greater insight into the scientific explanations of the properties, and behaviour, of soap bubbles and soap films. In common with Professor Boys's book the demonstrations here, particularly in Chapters 3 and 4, are simple to perform using household materials.

The first four chapters should be easily comprehended by anyone with an undergraduate training, or a strong sixth form background, as they contain mainly school mathematics. Chapter 1 contains a general introduction to the chemical, physical and mathematical concepts that will be developed in later chapters. Chapter 2 is concerned with the optical interference phenomena that are produced by soap films and their application to the study of the draining and thinning mechanisms present in the films. Chapters 3 and 4 explain how soap films can be used as an analogue computer to solve mathematical minimization problems in two and three dimensions. They also give some mathematical analysis and discussion.

Chapter 5 investigates the shape of liquid drops, bubbles, and the liquid surface in the vicinity of a solid surface, using the Laplace–Young equation. The last chapter, Chapter 6, contains a number of interesting properties and applications, such as the vibrational oscillations of soap film membranes and the application of soap films to the analogue solutions of the differential equations of Poisson and Laplace.

The proofs of the more difficult mathematical results have been included in the appendices at the end of the book. These proofs are discussed qualitatively in the main text so that readers, who do not have the appropriate mathematical background, will have no difficulty in following the discussions and explanations. At the end of the book there are references to books, scientific papers, educational films and other information about soap films and bubbles. These references are numbered and some of them referred to by superscripts in the text.

The author has visited numerous institutions in Britain and the United States giving lecture-demonstrations at all academic levels. Soap bubbles and soap films is a subject that can be appreciated by all ages. Primary school children can learn some simple geometrical properties and perform experiments for themselves. Older children will be able to appreciate some of the simpler scientific principles. At sixth form and undergraduate levels the more detailed explanations, presented in this book, can be given. For the researcher there are many questions that still remain to be resolved.

The shapes and interference colours produced by soap films and bubbles are visually attractive. So efforts have been made to produce photographs that show clearly the shapes and colours. The high standard of most of the coloured photographs, in the centre of the book, are due to Mr. Jim Styles of the Photographic Unit at the University of Kent at Canterbury. Most of the black and white photographs were taken by Miss Dorothy Finn, of the Physics Laboratory, and Mr. Jim Styles. I am most grateful for their time, care, and patience. Professor Karol J. Mysels, of the General Atomic Company, San Diego, California, in the United States, kindly gave permission for Plates 2.4, 2.5, and 2.6 to be reproduced from his book with Kozo Shinoda and Stanley Frankel, *Soap Films, Studies of their Thinning* and Professor J. Th. G. Overbeek of the State University of Utrect, The Netherlands, provided the interference Plates 2.2, 2.3, and 2.7 from his review article in volume II of *Chemistry, Physics and Applications of Surface Active Substances*. Professor J. F. Nye of Bristol University allowed me to reproduce bubble raft photographs, Plate 4.12, from his thesis. The acknowledgements to bodies who have given permission to reproduce copyright material are listed separately.

I would like to thank Professor Mysels for reading the original manuscript and making many useful comments and corrections. In addition I am grateful for the criticisms received from my colleagues Dr. C. R. Brown and Dr. James Bridge. Dr. A. L. Smith of the Unilever Research Laboratory, Port Sunlight, Merseyside, was most helpful and enabled me to seek expert opinions on a number of questions, particularly with Dr. Jaap Lucassen. The discussions and comments of Dr. Peter Richmond, of the same Laboratory, were much appreciated. In addition I would like to thank Professor R. Osserman of Stanford University in the United States for helpful discussions concerning the mathematics of minimum surfaces.

The hard work of typing the manuscript was carried out speedily and efficiently by Miss Naomi Nason, Miss Diane Jolly, and Mrs. Betty Jones, all of the Physics Laboratory at the University of Kent at Canterbury.

Finally, I would like to thank Mr. Mike Grover of Tieto Ltd., the publishers, for suggesting that I write this book and for his continual enthusiasm and help during the writing and printing of the book.

Cyril Isenberg
Physics Laboratory
University of Kent at Canterbury
England

CONTENTS

Preface

1 *GENERAL INTRODUCTION*

2 *DRAINING AND THINNING OF SOAP FILMS*

3 *THE MOTORWAY PROBLEM*

4 MINIMUM SURFACES IN THREE DIMENSIONS

5 THE LAPLACE–YOUNG EQUATION

6 ANALYTIC METHODS AND RESULTS, VIBRATIONAL MODES, AND FURTHER ANALOGUE METHODS

COLOUR AND BLACK AND WHITE
PLATES BETWEEN PAGES   52 and 53

APPENDIX

ACKNOWLEDGEMENTS

I am grateful to the following companies and institutions for permission to reproduce copyright figures and tables:

G. Bell & Sons Ltd. (Table 2.1)
Gordon and Breach, Science Publishers Ltd. (Plates 2.2, 2.3, 2.7)
Science Museum, London (Fig. 1.4)
Pergamon Press Ltd. (Plates 2.4, 2.5, 2.6)
The Royal Institution (Fig. 4.24)
Unilever Educational Publications (Plate 4.13)

1 GENERAL INTRODUCTION

1.1 Historical Review

The beauty of soap bubbles and soap films has a timeless appeal to young and old alike. It has been captured through the ages by such prominent painters as Murillo,[73] Chardin,[74] Hamilton,[75] Manet,[76] and Millais.[77]

The scientific study of liquid surfaces, which has led to our present knowledge of soap films and soap bubbles, is thought to date from the time of Leonardo da Vinci[102, 110]; a man of science and art. Since the fifteenth century researchers have carried out investigations in two distinct camps. In one camp there are the physical, chemical and biological scientists who have studied the macroscopic and molecular properties of surfaces with mutual benefit. The other camp contains mathematicians who have been concerned with problems that require the minimization of the surface area contained by a fixed boundary and related problems. A simple example of such a problem is the minimum area surface contained by a circle of wire. The solution to this problem is well known to be the disc contained by the wire.

In the nineteenth century the Belgian physicist Joseph Plateau[46, 47] showed that analogue solutions to the minimization problems could be produced by dipping wire frameworks into a bath of soap solution. After withdrawing a framework from the bath a soap film is formed in the frame, bounded by the edges of the framework, with a minimum area surface. All the minimum surfaces were found to have some common geometrical properties. This work rapidly attracted the attention of the mathematicians and has resulted in a fruitful interaction between the two camps. These experimental results have inspired mathematicians to look for new analytic methods to enable them to prove the existence of the geometric properties associated with minimum area surfaces and to solve the minimum area problems. However it is only relatively recently that important steps have been made in this direction, particularly the work of Jesse Douglas[86] and his contemporaries[21, 108] in the 1930's, and the recent work of mathematicians[78] in the United States.

Leonardo da Vinci, in the fifteenth century, studied the rise of a liquid up a capillary tube when it is inserted in a bath of liquid (Fig. 1.1). The liquid

1

rises above the level of the liquid surface in the bath. The first accurate quantitative observations of this phenomenon were made by Francis Hawksbee in 1709. Isaac Newton[42] was aware of this behaviour of capillary action and ascribed it to the attraction of the liquid by the tube. He and Robert Hooke[94] had also observed the colours and black spots in soap films and bubbles due to the interference of light produced by reflection from the surfaces of the film.

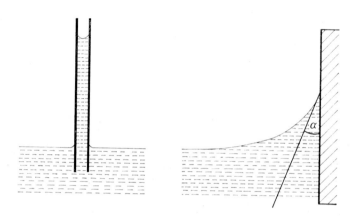

Fig. 1.1   The rise of a fluid in a capillary tube.

Fig. 1.2   The angle of contact, α, between a fluid and a solid.

The important concept of a tension in the surface of a fluid, known as *surface tension*, was introduced by J. A. von Segner in 1751. This is the force in the surface of a fluid acting on each side of a line of unit length drawn in the surface. Thomas Young[120] in his essay entitled *Cohesion of Fluids* in 1805 used the concept of surface tension to explain the rise of liquid in a capillary tube and introduced the concept of *angle of contact*.[43,119,120] This is the angle between a fluid surface and a solid surface (Fig. 1.2). His work contains the solution to a number of problems that were later solved, with the help of mathematical techniques, independently, by the Marquis de Laplace.[44] Young never received recognition by Laplace for his work and he never forgave Laplace for this. The result that is, perhaps, most widely known is due independently to Young and Laplace. It concerns the excess pressure, p, across a curved fluid surface, or a surface separating two fluids, at a point with principal radii of curvature R_1 and R_2. These are the maximum and minimum radii of curvature of the surface at the point. It is usually known as the Laplace–Young equation and states that

$$p = \sigma \left(\frac{1}{R_1} + \frac{1}{R_2} \right), \tag{1.1}$$

where σ is the surface tension at the fluid interface. We will discuss this result further in section 1.5 and use it frequently throughout the analysis of the shape of fluid surfaces, soap films and soap bubbles. Equation (1.1) was published by Laplace[44] in 1806 in his mammoth work *Mécanique céleste*.

Important theoretical contributions to the study of fluid surfaces were made by Carl Friedrich Gauss in 1830 and by Siméon Denis Poisson in 1831. Gauss re-derived the Laplace–Young equation by examining the energy of the fluid surface and obtained an expression for the angle of contact at the boundary. Poisson introduced the concept that the density of the fluid in the region of the surface was different from that of the bulk fluid.

Fig. 1.3 Joseph Antoine Ferdinand Plateau, 1801–1883.

Fig. 1.4 Charles Vernon Boys, 1855–1944.

Joseph Antoine Ferdinand Plateau[48, 79, 85, 104–5, 118] 1801–1883 (Fig. 1.3), devoted much of his life to the study of the surface properties of fluids despite being completely blind in the latter half of his life. However his early research interests were in the field of optics. In 1829 he performed an experiment in which he exposed his eyes to the sun's rays for 25 seconds. This caused permanent damage to his sight. His sight gradually deteriorated and by 1843, at the age of 42 years, he was completely blind. During the period in which he was partially sighted he became interested in the nature of the molecular forces present in the surface and the bulk of a fluid. It was during this period that he discovered the unusual surfaces formed by soap films contained by wire frames.

He published many scientific papers on this subject and other related subjects. These were summarized in a two volumed work entitled *Statique expérimentale et théorique des liquides soumis aux seules forces moléculaires*[46] (Experimental and Theoretical Investigations of the Equilibrium Properties of Liquids Resulting from their Molecular Forces) which was published in 1873. Despite being completely blind he continued his research at the University of Gent, in Belgium, with the help of his colleagues and family. In 1844 he was made a research professor, without any teaching duties. In later life he received many honours for his contributions to the study of fluids and soap film surfaces, which have so greatly influenced the work of mathematicians in their study of minimum area surfaces.

In the latter half of the nineteenth century Josiah Willard Gibbs,[51] who is well known for his theoretical contributions to the study of statistical mechanics and thermodynamics, observed the draining and thinning of soap films. Some of these observations are reported in his paper entitled *Equilibrium of Heterogeneous Substances*.[49, 52]

The great popularizer of the properties of soap films and soap bubbles at the era around the turn of this century was Charles Vernon Boys (Fig. 1.4). He gave numerous lecture-demonstrations which delighted everyone, no matter their age or academic background, and his book *Soap Bubbles and the forces which mould them*[1] has been popular with young people since its publication in 1890.

Another popularizer of soap bubbles at the beginning of this century was the biologist D'Arcy Wentworth Thompson who drew attention to the similarities between the shapes of soap bubbles and the shapes that occur in living organisms. This is discussed in his classic book *On Growth and Form*.[37]

During the First World War Sir James Dewar,[53] who is primarily known for his work in the field of low temperature physics and for the creation of the Dewar flask, was unable to continue his researches into low-temperature phenomena. So he investigated the draining and stability of soap films. His assistant and colleague during this period was Mr. A. S. C. Lawrence who summarized their joint work in his book, *Soap Films, a study of molecular individuality*.[8]

A large chemical industry has grown up which is dependent on the science of surface phenomena. Many of the products produced by these industries have been developed by applying our understanding of surface properties such as wetting, dyeing, foaming, coalescing and emulsification. The study of soap films and soap bubbles forms an important part of the science of surface phenomena and consequently a large programme of research in this field is being carried out in industry and universities.

In recent years biochemists[27, 38] have been studying, with increasing interest, biological membranes present in animal and plant cells. They are composed of *lipid* molecules that are similar in structure and behaviour to soap molecules. A study of soap films can provide an insight into the function and properties of these lipid molecules and lipid membranes.

What have been the significant historical developments in the mathematics of minimum area surfaces? John Bernoulli and his student Leonhard Euler were amongst the earliest workers to apply the methods of the calculus to the solution of these problems, thus laying the foundations for a new branch of the calculus, the *Calculus of Variations*.[55] In a comprehensive work published in 1744 Euler[41] derived his well know equation for the determination of minimum area surfaces and other *variational* problems that require the examination of a sequence of varied surfaces. The equation, in its simplest one dimensional form, is

$$\frac{d}{dx}\left(\frac{\partial f}{\partial y_x}\right) - \frac{\partial f}{\partial y} = 0, \tag{1.2}$$

where $y = y(x)$, $y_x = (dy/dx)$, and $f = f(x, y, y_x)$.

The derivation of this equation was based on a geometric-analytic method. Euler successfully applied his equation to the celebrated problem of the determination of the minimum area surface contained by two parallel co-axial rings, arranged perpendicular to their common axis. This was found to be a catenary of revolution, or catenoid, providing the rings are sufficiently close together.

Joseph Louis Lagrange[101] was attracted by Euler's work and reformulated it using purely analytic methods. Equation (1.2) has since become known as the Euler–Lagrange equation. The solution to the minimum area surface contained by two coaxial rings remains one of the few analytic solutions available in this field.

It was Joseph Plateau's experiments with soap films that provided mathematicians with renewed motivation to investigate the problems of minimum area surfaces. Some of these beautiful analogue solutions are examined in Chapter 4 with coloured plate illustrations. Although substantial efforts were made to obtain analytic solutions to these problems it was not until the 1930's that significant progress was made by mathematicians such as Jesse Douglas,[86] who obtained some general solutions, and Tibor Radó.[24]

More recently new mathematical techniques[108] of differential geometry have been developed which required the use of *currents, varifolds,* and *geometric measure theory*. They are being applied with some success to the *Plateau problem*; the determination of the minimum area contained by a boundary.

1.2 Surface Tension

The molecules near the surface of a pure fluid have a different environment from those in the interior of the fluid (Fig. 1.5(a)). A molecule in the bulk of the fluid will experience forces in all directions due to the surrounding molecules. The resultant force on such a molecule averaged over a macro-scopic time, a time which is much longer than that between two collisions, will be zero. Molecules near the surface of the fluid will experience a weaker force, from the gaseous region above the surface, than they would experience if the gaseous region was replaced by fluid, as the density of the gaseous region is considerably smaller than that of the bulk fluid. Consequently such mol-ecules will experience, on average, a force pulling them back into the bulk of the fluid as indicated in Fig. 1.5(a) This force will have the effect of reducing the area of the surface providing the surface is free to change its shape, as in the case of a water droplet which always takes up a spherical shape. It will also have the effect of reducing the density of the fluid in the region of the surface. Figure 1.5(b) shows the variation in density, ρ, of a typical fluid as a function of distance, r, measured perpendicular to the surface from the bulk fluid to the gaseous region.

A soap solution consists of soap molecules and water molecules. Each soap molecule is formed from the metal salt of a long chain fatty acid molecule and becomes ionized in solution. For example in the case of the soap, sodium stearate, the sodium ions have a positive charge and are dispersed throughout

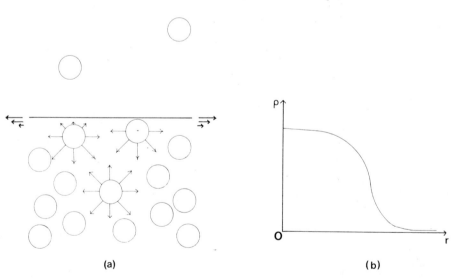

(a)   (b)

Fig. 1.5(a)   Molecular forces experienced by a molecule in the bulk, and at the surface, of a fluid.

Fig. 1.5(b)   The density, ρ, as a function of distance, r, across the surface of a fluid.

the solution. The negatively charged stearate ions near the surface, which consist of a negatively charged 'head' and a hydrocarbon 'tail', experience an average force towards the surface. Some of them will accumulate in a mono-layer at the surface; they are *adsorbed* on to the surface. These surface, or *surfactant*, ions have their negatively charged 'heads' in the surface with the hydrocarbon 'tails' out of the surface (Fig. 1.6). This is the energetically favoured configuration for the ions in the surface. The hydrocarbon chains are squeezed out of the surface by the water molecules. Some of the stearate ions will remain in the bulk fluid. This general structure, that results from the adsorption of ions onto the surface, will also be true of the artificial soaps which are also synthetic detergents, and lipid systems that will be discussed in section 1.6.

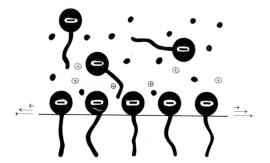

Fig. 1.6 Negative soap ions in the
surface of a soap solution.

Fig. 1.7 Surface tension is perpendicular to any line drawn
in the surface.

An important consequence of the variation in the environment of the molecules in the region of the surface of the fluid, in both pure fluids and soap solutions, is the presence of a macroscopic surface force localized within about one atomic thickness of the surface. For most purposes it is justifiable to consider this as a surface tension, that is a force per unit length, σ, in a 'mem-brane' of negligible thickness at the surface of the fluid.

A soap film consists of two such surfaces separated by a thin layer of fluid, which may vary in thickness from 2×10^5 Å to 50 Å (10 Å = 1 nm). This ranges from 50 times the wavelength of visible light to a few atomic distances. The largest thickness will occur immediately after the formation of a film. Once the film has formed it will commence to thin. Each surface is composed of soap ions which are separated, largely, by water molecules. The surplus water will drain away from the film by various draining processes. These pro-cesses will be described in Chapter 2. The thickness of the film will decrease until a final equilibrium thickness is reached, providing the film does not rupture. Both surfaces of the film will have a surface tension associated with them.

The surface of any fluid will be in a state of *uniform tension* if the fluid surface satisfies the following conditions:

(a) The surface tension must be perpendicular to any line drawn in the surface and have the same magnitude for all directions of the line (Fig. 1.7).

(b) The surface tension must have the same value at all points on the surface.

In the case of a soap film, with two surfaces, it is convenient to introduce the concept of film tension, σ_f, which is the force per unit length of film and is equal to twice the surface tension. For thick films the surface tension will be equal to that at the surface of a bath of soap solution. However for very thin films the value will differ from that for the surface of a bath of soap solution.

The two conditions, (a) and (b), are satisfied by most fluids. For a soap film, however, it is possible that (b) will only be satisfied approximately. For example consider the equilibrium of a section of a vertical soap film of thickness t, width l, and height h (Fig. 1.8). Let the surface tension at the bottom of the section of film be σ_0 and that at height h, σ_h. The vertical force at the top of the film is $2l\sigma_h$, the factor of 2 arises because the film has two surfaces. This force balances the force at the bottom of the film, $2l\sigma_0$, plus the weight of the film mg, where m is the mass of the film. If ρ is the density of the fluid in the film then $m = tlh\rho$. For equilibrium,

$$2l\sigma_h = 2l\sigma_0 + tlh\rho g. \tag{1.3}$$

Hence

$$\frac{\sigma_h - \sigma_0}{\sigma_0} = \frac{th\rho g}{2\sigma_0}. \tag{1.4}$$

A thick film with a thickness of one micron (10^{-4} cm), a height of 10 cm, and $\sigma_0 = 30$ dynes per cm has, from Eq. (1.4), a variation in surface tension, σ, of about 1.5 per cent. In the thinnest films, typically 60 Å thick, this variation is reduced to 0.01 per cent.

It has been assumed that the fluid, or soap film, is at constant temperature in thermodynamic equilibrium. Under these conditions it is found that the surface tension, σ, of a fluid surface depends only on the temperature. This is known as the static surface tension. The surface tension will differ from the static value if the fluid, or soap film, is not in thermodynamic equilibrium. An example of a common non-equilibrium situation occurs in a jet of water issuing from a pipe (Fig. 1.9). The molecules at the surface of the water are not in thermodynamic equilibrium. The environment of the molecules at, or near, the surface will differ from that of a fluid in thermodynamic equilibrium. Consequently the surface tension will differ from that in the static case. This is known as the dynamic surface tension. The dynamic surface tension will vary from one non-equilibrium situation to another and will also depend on the time that has elapsed since the formation of the surface.

Surface tension has been defined as the force per unit length in a liquid-vapour interface. The concept can be extended to two phases of different fluids providing they do not mix; immiscible fluids. The surface tension between two liquid phases is called the *interfacial tension*, and that between a solid and a liquid the *adhesion tension*. There will also be a surface tension at a solid-gas interface.

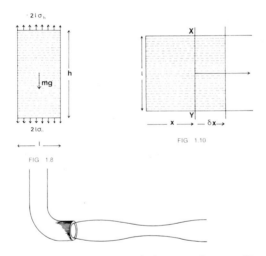

Fig. 1.8   Forces acting on a vertical rectangular soap film.

Fig. 1.9   The surface of a jet of water has a *dynamic* surface tension.

Fig. 1.10   A soap film contained by a rectangular frame with a moveable side XY.

The surface tension of a fluid varies with temperature and differs in behaviour for associated and unassociated liquids. An unassociated liquid is one which consists only of individual molecules. An associated liquid consists of groups of attached molecules. Examples of associated liquids are water and formic acid, and examples of unassociated liquids are benzene and carbon tetrachloride. The groups of molecules in associated liquids break up as the temperature rises. This produces a different variation of the surface tension with temperature from that for unassociated liquids.

The surface tension of an unassociated liquid in thermodynamic equilibrium with its vapour has been shown empirically to be of the form,

$$\sigma = \sigma_0 \left(1 - \frac{T}{T_c}\right)^n, \tag{1.5}$$

where T is the absolute temperature, T_c is the critical temperature at which σ vanishes, and σ_0 and n are constants for each liquid. A typical value of n is 1.2. It is seen, from Eq. (1.5), that σ decreases with increasing temperature, becoming zero at $T = T_c$.

1.3   Energy Considerations

In order to obtain an expression for the energy of a soap film, or liquid surface, let us consider a soap film contained by a rectangular wire frame of width l (Fig. 1.10), with one side of the frame, XY, free to move in the direction perpendicular to XY. If this side is initially at a distance x from the parallel fixed side and undergoes a further displacement δx, maintaining the temperature of the film constant, the work done against the film tension force is

$$(\sigma_f l)\delta x = \sigma_f \delta A, \tag{1.6}$$

where σ_f is the film tension and δA is the increase in the area of the film. For a thick film $\sigma_f = 2\sigma$, where σ is the surface tension of the surface of soap solution in a bath. For thin films σ_f will differ from this value but will be twice the magnitude of the surface tension of the film. From Eq. (1.6) σ_f is seen to be the energy gained per unit increase in area. Thus the total energy necessary to increase the area of a soap film from zero area to A is given by F, where

$$F = \int_0^A \sigma_f dA. \tag{1.7}$$

F is called the *free energy* at constant temperature. It is the energy, at constant temperature, available to perform mechanical work. If the concentration of surfactant molecules in the soap film is sufficiently great σ_f is found to be independent of area, A. In this case σ_f is constant and result (1.7) becomes

$$F = \sigma_f A. \tag{1.8}$$

The free energy of the soap film is thus proportional to the area of the film under the conditions of constant σ_f. This important result is used to obtain analogue solutions to mathematical problems in Chapters 3, 4, and 6. If the concentration of surfactant molecules is not sufficiently great σ_f will depend on the area of the film, A, and consequently the free energy will be given by the integral expression in (1.7). The results (1.6), (1.7), and (1.8) also apply to the free energy of the surface of a fluid providing σ_f is replaced by the surface tension, σ, as a fluid has only one surface.

When a soap film is in stable equilibrium any small change in its area, A, will produce a corresponding change in its free energy, F, providing σ_f remains constant, Eq. (1.8). As F is minimized when the film is in stable equilibrium, A will be minimized. For example a soap film with constant film tension, σ_f, contained by a circular ring of wire is known, when it is in equilibrium, to have the shape of the disc bounded by the wire ring. This configuration is known to have the property of minimum area. Any other surface, bounded by the ring, will have a greater area.

Another example of the minimum area property is obtained if a loop of cotton thread is joined to the circular ring and a soap film is formed in the ring. It will rest loosely in the plane of the soap film (Plate 1.1(a)). When the soap film inside the loop of thread is broken the surrounding soap film, contained between the loop of thread and the circular ring, will take up the surface of minimum area. This will occur when the area of the hole inside the loop of thread is a maximum. The maximum area of the hole can be shown mathematically to occur, for a fixed length of thread, when it forms a circle. This is seen to occur in Plate 1.1.(b).

A further example is provided if a length of thread has both its ends tied to two points on the ring so that the thread hangs limply between the attachment points when a disc of soap film is formed in the ring. Now if the soap film on one side of the thread is broken, the remaining soap film will take up the minimum area contained by the thread and remaining part of the circumference of the ring bounding the soap film. The shape taken up by the thread, that minimizes the enclosed area of soap film, is shown in Plate 1.1(c). It is an arc of a circle providing the length of the thread is less than the arc of the wire circle bounding the film. This is also known mathematically to provide the surface of minimum area.

The minimum area property of soap films can also be used to illustrate the minimum area surface contained by two coaxial rings which was shown by Euler to be a catenoid surface, providing the rings are sufficiently together. The result of dipping two rings into a bath of soap solution and forming the simplest minimum area soap film surface joining the two coaxial rings is shown in Fig. 1.11. It is indeed a catenoid surface.

A soap film, such as that in Fig. 1.10, when expanded in area, at constant temperature, will absorb heat from its surroundings in order to maintain the temperature of the film constant. If the film is expanded rapidly it will not have time to absorb heat energy from its surroundings and will consequently cool. Thus, in addition to the energy gained by the film (Fig. 1.10), from work done on the system, there is an absorption of heat energy, δQ. If ϵ_f is the intrinsic surface energy per unit area of the film, that is the total energy ab-

Fig. 1.11   The soap film joining two rings has the shape of a catenoid.

sorbed per unit area by the film during a displacement δx, then conservation of energy requires that

$$\epsilon_f l \delta x = \sigma_f l \delta x + \delta Q. \tag{1.9}$$

That is

$$\epsilon_f = \sigma_f + \frac{1}{l} \frac{\delta Q}{\delta x}. \tag{1.10}$$

Now according to thermodynamics[28]

$$\frac{1}{l} \frac{\delta Q}{\delta x} = -T \left(\frac{\partial \sigma_f}{\partial T} \right)_x. \tag{1.11}$$

(This condition is obtained by ensuring that the total free energy of the soap film at *any temperature*, F, given by $F = U - TS$, where U is the internal energy and S is the entropy, is a perfect differential[12, 13, 28].)

So (1.10) becomes, in general,

$$\epsilon_f = \sigma_f - T\left(\frac{\partial \sigma_f}{\partial T}\right)_A. \qquad (1.12)$$

A similar derivation applies for the intrinsic surface energy per unit area of a fluid, ϵ, providing σ_f is replaced by σ, the surface tension of the fluid surface. Thus in general for a fluid surface

$$\epsilon = \sigma - T\left(\frac{\partial \sigma}{\partial T}\right)_A. \qquad (1.13)$$

Figure 1.12 shows the typical variation of σ and ϵ with T. For unassociated liquids Eq. (1.5) gives the empirical relationship between σ and T from which

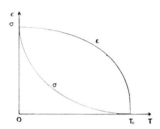

Fig. 1.12   The variation of σ and ϵ with temperature, T.

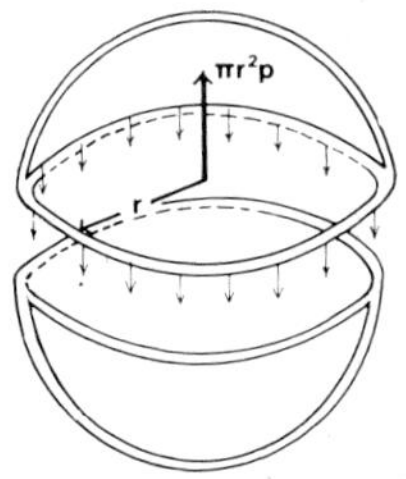

Fig. 1.13   The forces acting on a hemispherical section of a bubble.

ϵ can be calculated by using Eq. (1.13). It will be observed that both σ and ϵ are zero at $T = T_c$, and at $T = 0$ both σ and ϵ are equal. These limiting results also hold for ϵ_f and σ_f. ϵ is always greater than σ at intermediate temperatures as $-T(\partial \sigma / \partial T)_A$, the heat absorbed per unit area by the surface, or film, is always positive.

1.4   Equilibrium of Soap Bubbles

A soap bubble consists of a thin spherical shell which is composed of water and soap ions. Inside the shell is air or gas at a greater pressure than the external atmospheric pressure. By examining the forces acting on a hemispherical section of a soap bubble (Fig. 1.13), we can relate the excess pressure inside the bubble over the surrounding pressure, p, to the radius of the bubble, r, and the film tension, σ_f.

The forces acting on a hemispherical section of a soap bubble are the film tension force, $2\pi r \sigma_f$, pulling vertically downwards (Fig. 1.13), and the force due to the excess pressure, p, acting vertically upwards on the hemispherical section of bubble, $\pi r^2 p$. The weight of the hemispherical shell of fluid can be neglected compared with either of these forces. As the hemispherical section is in equilibrium these two forces must balance, thus

$$\pi r^2 p = 2\pi r \sigma_f. \tag{1.14}$$

Hence

$$p = \frac{2\sigma_f}{r}. \tag{1.15}$$

This result is in agreement with the result obtained by the application of the Laplace–Young equation, Eq. (1.1). For a spherical bubble the principal radii of curvature, R_1 and R_2, are equal to its radius, that is $R_1 = R_2 = r$. So the excess pressure across each surface of the bubbles is, from Eq. (1.1),

$$\frac{2\sigma}{r}, \tag{1.16}$$

where σ is the surface tension of the film. As the bubble has two surfaces a double application of the Laplace–Young equation gives the excess pressure, p, inside the bubble as

$$p = \frac{4\sigma}{r}. \tag{1.17}$$

As $\sigma_f = 2\sigma$, this result is in agreement with the result (1.15). For a spherical liquid drop, having only one surface, with surface tension σ, the analysis using the equilibrium of a hemispherical section, or that using the Laplace–Young equation, gives an excess pressure inside the drop of

$$p = \frac{2\sigma}{r}. \tag{1.18}$$

Thus bubbles and drops with small radii have large excess pressures and visa versa. A plane surface of soap film or fluid can be considered as part of a spherical surface of large radius and so the excess pressure difference across the surface is zero.

It is instructive to examine the equilibrium of two bubbles, A and B, both of radius r connected by a pipe, as indicated in Fig. 1.14, assuming that the concentration of the surfactant soap ions, the surface ions, is sufficiently great to ensure that σ_f remains constant for variations in r. The pressure in the two bubbles will initially be equal. Small fluctuations in the air pressure will cause a small quantity of air to be transferred from, say, A to B. The radius of B will increase and so the excess pressure in B will decrease, Eq. (1.15). The converse will be true for bubble A; the radius of A will decrease with a consequent increase in the excess pressure. Thus the pressure difference between A and B will continue to increase. The bubble B will grow whilst bubble A shrinks. The system of bubbles is unstable, and the radius of curvature of the smaller bubble will continue to decrease until it becomes equal to the diameter of the pipe to which it is connected (Fig. 1.14). Beyond this state the radius of curvature of the bubble will increase as it now forms the minor spherical cap of a sphere bounded by the orifice of the pipe. The excess pressure will thus begin to decrease in bubble A. Eventually a radius of curvature of A will be reached that is equal to that of B, and stable equilibrium will be attained. The broken curves in Fig. 1.14 indicates this final state.

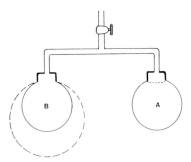

Fig. 1.14 Two bubbles in equilibrium; the full curves represent the initial unstable position, the broken curves represent the final stable equilibrium position.

In order for two bubbles, A and B, to be in stable equilibrium, a small transfer of air from A to B must be associated with a pressure increase in B, so that air flows back to restore the equilibrium and visa versa. This condition is satisfied by the final equilibrium state of the two bubbles examined above under the constraints imposed by the radii of curvature of the exit outlets of the pipe. Thus for stable equilibrium we require that the excess pressure, p, and radius, r, of each bubble satisfy,

$$\frac{dp}{dr} > 0. \qquad (1.19)$$

It has been assumed that σ_f remains constant in investigating the behaviour of the bubbles. For soap films with low surfactant concentrations σ_f will not remain constant. As the radius of the bubble decreases more soap ions will be adsorbed into the surface, which will decrease σ_f. By increasing the radius of the bubble the film will become more water-like, as the surface density of surfactant ions will decrease. Hence σ_f will increase as water has a higher surface tension than soap solution. Under these conditions a bubble will be in stable equilibrium if condition (1.19) is satisfied. That is, from (1.15), if

$$\frac{d}{dr}\left(\frac{2\sigma_f}{r}\right) > 0. \tag{1.20}$$

Differentiating (1.20), in which σ_f now depends on r, gives

$$\frac{2}{r^2}\left(-\sigma_f + r\frac{d\sigma_f}{dr}\right) > 0, \tag{1.21}$$

or

$$\frac{d\sigma_f}{d\log r} > \sigma_f. \tag{1.22}$$

The area of the bubble $A = 4\pi r^2$. So (1.22) can be written as,

$$2\frac{d\sigma_f}{d\log A} > \sigma_f. \tag{1.23}$$

The quantity

$$\gamma = \frac{d\sigma_f}{d\log A} \tag{1.24}$$

was first introduced by Willard Gibbs and is called the *surface elasticity* of the film. The condition for two or more bubbles to co-exist in stable equilibrium with the same excess pressure is, from (1.23) and (1.24),

$$2\gamma > \sigma_f. \tag{1.25}$$

Similarly, the surface elasticity of the surface of a soap solution can be defined by replacing σ_f in (1.24) by σ.

1.5 Pressure Difference across a Curved Fluid Surface

The most general surface separating a liquid and a gas or two immiscible liquids will have, at every point on the surface, a maximum and a minimum radius of curvature, R_1 and R_2 respectively. These are the principal radii of curvature and occur in planes that are perpendicular to each other, and are both perpendicular to the tangent plane to the surface. It was mentioned earlier that the Laplace–Young equation relates the excess pressure across the surface at any point to these radii of curvature at the point by

$$p = \sigma \left(\frac{1}{R_1} + \frac{1}{R_2} \right), \tag{1.26}$$

where σ is the surface tension for the liquid–gas interface, and the interfacial tension for a fluid–fluid interface.

It was pointed out in the last section that result (1.26) reduces to result (1.18) for a spherical drop of liquid and to (1.15) for a spherical soap bubble. For the case of the catenoid soap film surface with zero excess pressure across the surface bounded by two coaxial rings, that was discussed in section 1.3, Eq. (1.26) gives

$$\frac{1}{R_1} + \frac{1}{R_2} = 0. \tag{1.27}$$

This equation is of general validity for any soap film surface with zero excess pressure across it.

It is shown in Appendix IV, for a restricted class of problems, that Eq. (1.27) is the mathematical condition for a surface to have minimum area. This mathematical minimum area property can be shown to hold generally for any surface that has zero excess pressure across it and hence satisfies Eq. (1.27). We can verify this result on physical grounds by considering the energy of a soap film, as in section 1.3, and showing that the condition for stable equilibrium of a soap film requires that the area of the film be minimized providing that σ_f is constant.

1.6 Molecular Structure of Soap Films

Soap solutions have the remarkable property of forming stable bubbles and films. This property is a consequence of the surface structure of the soap solution and the soap film. The surface of a bath of soap solution and a soap film consists of a monomolecular layer of *amphipathic*† ions. These are ions

---

† Amphipathic from the Greek word ἀμφιπάθεια meaning 'both loving'.

that have two dissimilar parts. One part is hydrophilic, water loving, which means that it likes to be surrounded by water as a result of the attraction between the hydrophilic part of the ion and the water molecules. The other part of the ion is hydrophobic, water hating, which means that it has a dislike for a water environment that results from the relative magnitudes of the attractive forces of the water molecules for each other and the attractive forces of the hydrocarbon 'tails' for each other compared with the water-'tail' attractions.

Ordinary soap is a sodium salt of a fatty acid. It has a hydrophilic polar carboxyl 'head' and a long hydrophobic 'tail' consisting of a hydrocarbon chain. The hydrophilic 'heads' of soap ions that are adsorbed into the surface are surrounded by water molecules and ions. Their hydrophobic, water hating, 'tails' are directed out of the surface. The amphipathic ions at the surface of the soap solution or film can orient themselves so that their 'heads' lie in the water and their 'tails' remain out of water (Fig. 1.15). The bulk fluid in the soap solution or film will also contain, in addition to water molecules and metal ions, some amphipathic ions. An example of a soap molecule is sodium stearate, $C_{17}H_{35}COO^-Na^+$. The polar, hydrophilic, part is the COO^- group and the, hydrophobic, hydrocarbon chain is $C_{17}H_{35}$. The positive metallic sodium ions are dispersed throughout the bulk fluid in the soap solution or film. A typical surfactant amphipathic ion occupies an area of $40Å^2$ and has a length of about $30Å$. Fig. 1.15(a) shows the surface of a bath of soap solution and Fig. 1.15(b) shows a soap film consisting of two surfaces of amphipathic ions. The amphipathic soap ions are adsorbed into the surface of the soap solution as a result of the hydrophobic interactions near the surface of the soap solution between the hydrocarbon chain of the ion and the water molecules. The hydrocarbon chains emerge from the liquid surface by a process in which the water molecules squeeze the hydrocarbon chains out of the surface by the attractive interactions.

Some of the amphipathic ions will exist in the bulk fluid. As the concentration of these ions is increased, by the addition of soap molecules, a critical concentration will be reached at which the ions in the bulk fluid will find it energetically favourable to form clusters. The clusters of ions will vary in size but usually contain at least fifty ions. These groups of ions contain the hydrophilic 'heads' on the outside of the cluster and the hydrophobic 'tails' directed to the centre of the cluster. In this way the hydrophobic 'tails' are excluded from the presence of water molecules whilst the hydrophilic 'heads' are in contact with the water molecules. These groups of ions are known as micelles[89] (Fig. 1.15(c)). The groups of ions will be highly charged and will attract metal ions of the opposite charge. Consequently the net charge, resulting from the negative charge on the micelle and the positive metal ions in the neighbourhood of the micelle, will be appreciably less than that associated with the sum

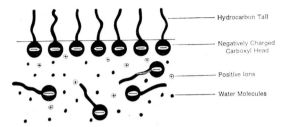

Fig. 1.15(a)   Surface structure of a soap solution.

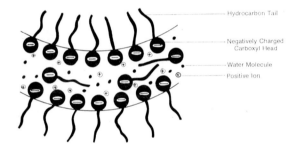

Fig. 1.15(b)   Structure of a soap film.

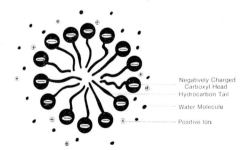

Fig. 1.15(c)   A micelle.

of the individual amphipathic ions comprising the micelle. The formation of micelles will only occur if the concentration of the soap solution exceeds a minimum value. This minimum value is known as the critical miscellization concentration, which is abbreviated by c.m.c. The density of surfactant ions in the surface of a soap solution varies appreciably with the concentration of soap solution in the range zero to about 0.1 c.m.c., but remains almost constant beyond this concentration. In this latter range the surface density is typically one amphipathic ion per 50Å^2.

The surface tension of soap solution is typically about one third that of water, which has a surface tension of 72.25 dynes per cm at 20 °C, and varies with the concentration, c, of the soap solution. Fig. 1.16 shows the variation of the surface tension, σ, with concentration, c. It has the value for distilled water at $c = 0$ and decreases monotonically until it reaches its smallest value at the c.m.c., as indicated in Fig. 1.16.

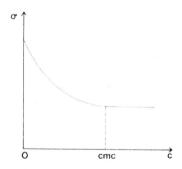

Fig. 1.16   The variation of surface tension, σ, with concentration, c.

The stability of soap films is determined by the amphipathic ions in the surface. If a soap film is perturbed from equilibrium so that the area of an element of film increases, the surface density of amphipathic ions will decrease. That is, the number of ions per unit area will diminish and consequently the surface will behave more like the surface of water. Hence the surface tension of the surface element will increase because the surface tension of water is greater than that of soap solution. This increased force in the region of increased area will restore the surface to its former equilibrium configuration. This stabilizing effect was first observed by Marangoni and is known as the *Marangoni effect*.

The Marangoni effect results from the variation of the surface tension, σ, with a change in the area of an element of the surface. So one might expect the effect to be related to the Gibbs elasticity of the surface, $d\sigma/d \log A$, introduced in section 1.4 for a system in thermodynamic equilibrium. However the Marangoni effect occurs under non-equilibrium conditions in which amphipathic ions from the bulk fluid are being adsorbed into the surface during and after the deformation of the surface. No simple relationship has been found between the Marangoni effect and Gibbs elasticity although there is a link between the two effects.

1.7 Soap Solutions for Films and Bubbles

Natural soaps are the sodium and potassium salts of fatty acids. They are produced as a result of the interaction of a caustic alkali and a fat. A typical animal or vegetable fatty acid molecule consists of a long hydrocarbon chain with a terminal carboxyl group. For example in the case of sodium stearate, which was introduced as an example in section 1.6, the fatty acid is stearic acid and it has the chemical formula $CH_3(CH_2)_{16}COOH$. The chemical reaction of stearic acid and sodium hydroxide results in the soap sodium stearate, $C_{17}H_{35}COO^-Na^+$. The commonly occurring fats are lauric, myristic, palmitic, stearic, oleic, etc. Common bars of washing soap will consist of a number of pure soaps.

In the last section it was explained that soap molecules ionize in aqueous solution. They form anionic surfactants in which the hydrophilic negatively charged carboxyl 'heads' lie in the surface and the neutral hydrophobic hydrocarbon 'tails', to which each 'head' is attached, point out of the surface. It is also possible to produce other synthetic surfactants, being also synthetic detergents, which are similar in structure to the soaps. The molecules are amphipathic and can be anionic, cationic, and nonionic. There is another category called zwitterion which has both anionic and cationic characteristics. The anionic surfactants are produced from alkylaryl, and alkyl, sulphonates and alcohol sulphates and are used primarily as detergents. Nonionic detergents, which include polyether esters, have become important in producing detergents that do not foam for use in washing machines. The cationic detergents include the quaternary ammonium salts.

Recently biochemists have come to understand the importance of cell membranes in animal and vegetable cells. These membranes consist of molecules called *lipids*[27, 38] which are similar in character to soap molecules. They form surfactant monolayers and micelles in solution (Fig. 1.17(a) and (b)), and the cell membranes are constructed from bilayers of ions (Fig. 1.17(c)). The lipid membranes also form closed 'cellular' regions which contain fluid, known as *visicles*, which have important biological functions. The lipids are amphipathic water-insoluble organic substances, found in all cell membranes, which are extractable by non-polar solvents such as chloroform, ether and benzene.

Phospholipids, which are common in mammalian membranes, contain a phosphoric acid ester. A simple example is phosphatidic acid which replaces the fatty acids associated with soaps. The molecules of these acids contain two hydrocarbon hydrophobic chains as indicated in the lipid–water systems of Fig. 1.17. The similarity between the structure and behaviour of lipids and soaps has led to a resurgence of interest in the properties of soap molecules.

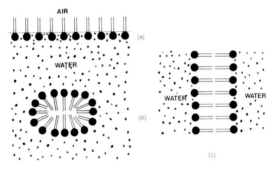

Fig. 1.17   Lipid-water systems. (a) A monolayer. (b) A micelle. (c) A bilayer.

The lifetime of pure soap films and bubbles is sensitive to the presence of impurities, dust particles, excess caustic alkali or excess fat. Consequently special care is necessary in the preparation of pure soap solutions and the subsequent formation of films and bubbles. This is, however, not true of the synthetic detergents. Only distilled water should be used for soap films with the longest life.

The stability and lifetime of films and bubbles are affected by the evaporation of water from the surface, the humidity of the surroundings, air currents, shocks and vibrations. Carbon dioxide in the atmosphere also diminishes the life of soap films. These factors can be eliminated by controlling the environment of the film or bubble. For example the bubbles and films can be produced in a closed enclosure with an atmosphere that has a saturated humidity, to prevent evaporation, and is free from shocks, vibrations, air currents and foreign gases. The lifetime of the film can be increased by the addition of glycerine to the solution. This has the effect of significantly reducing the evaporation and stabilizes the film. The pure soap film in a controlled environment should last indefinitely. This is not true for all bubbles. Bubbles contain gas at an excess pressure, above the environmental pressure. This gas will effuse through the bubble once the bubble has thinned significantly, with the result that the diameter of the bubble will decrease with time. The effect is greatest for the smallest bubbles which contain gas at the greatest excess pressure. Sir James Dewar[8] produced a bubble of diameter 32 cm in a controlled environment which he kept for 108 days. During this period the diameter decreased by a few centimetres due to effusion. He also produced a disc of soap film, 19 cm in diameter, which he kept for over three years.

Since the work of Joseph Plateau many workers have produced recipes for long lasting soap films and bubbles. A summary of these recipes is to be found in the book by J. J. Bikerman[4] and in the original works of Plateau,[46, 47] Boys,[1] Dewar,[53] Lawrence,[8] etc. It is often useful for demonstration purposes to produce a soap solution that will form films and bubbles with long life-times in the open air, without special precautions. The recipes mentioned above have been widely used for this purpose. More recently Cook[84] and Kuehner[100] have described solutions that are particularly good for this purpose. Kuehner has obtained 20 cm diameter bubbles that last for 102 minutes in the open air.

Many of the demonstrations described in this book do not require a specially prepared soap solution with the property of long lifetimes. A suitable soap solution can be prepared using tap water plus 1–2 per cent of any washing-up liquid (liquid detergent), for example Sunlight liquid or Fairy liquid. Such a solution will produce films with lifetimes of about 15 seconds.† The solution should always be thoroughly stirred before use and any bubbles that are formed on the surface of the solution should be removed. This solution can be made up quickly in small or large quantities. Some of the large scale demonstrations, to be discussed in Chapter 4, use a plastic dustbin containing 20 gallons of tap water and one small household container of washing-up liquid. Adding glycerine to the soap solution will increase the lifetime of soap films and bubbles. This increases from 15 seconds to minutes for solutions with 5 per cent glycerine and increases to hours for solutions containing 50 per cent glycerine. The solution should always be thoroughly stirred before use, if possible using an electrically driven stirrer. If such films are produced in an enclosure they will last for days or weeks. The soap bubble solution that can be purchased for bubble blowing is capable of producing bubbles that last for minutes.

There are many liquids that are capable of forming bubbles and films which are not soaps or synthetic detergents. They do not have the molecular surface structure of soap films which results from the adsorption of the soap ions into the surface of the solution. Examples of such liquids are plastic solutions,

---

† The lifetime is significantly reduced in a *dry* environment with a relative humidity of less than 50 per cent. In such conditions it is necessary to increase the relative humidity, in order to produce films and bubbles with lifetimes of 15 seconds, by using a humidifier or by simply evaporating water into the air by means of a cloth dipping into a trough of water.

liquid glass, and saponin. They produce rubber-like membranes. A saponin solution, which can be prepared from horse chestnuts, looks very much like soap solution and produces bubbles. However the difference appears once one withdraws air from a saponin bubble. It no longer retains its spherical shape and the skin will pucker. However if left for some time it will regain its spherical shape.

1.8   Interference Phenomena

Soap films and soap bubbles produce monochromatic interference fringes when exposed to monochromatic light and coloured fringes on exposure to white light. These interference phenomena occur when the thickness of the soap film is comparable to the wavelength of visible light. For example if a rectangular frame is withdrawn vertically from a bath of soap solution, a system of horizontal fringes is produced by light reflected from the film (Plates 2.1 and 2.2). This is due to the wedge-shaped profile of the soap film formed by the two faces of the soap film (Fig. 1.18(a)). The angle of the 'wedge' is too small to be seen by the eye but it can be detected by the interference pattern produced by visible light.

A ray of monochromatic light striking the film is split into two rays (Fig. 1.18(a)). One of these rays results from reflection at the film surface and constitutes about 4 per cent of the incident intensity. The other ray, which

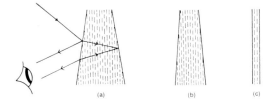

Fig. 1.18(a)   Interference of light produced by a vertical soap film after withdrawal from a bath of soap solution. (b) The film sometime later. (c) The final equilibrium film.

constitutes about 96 per cent of the incident intensity, is refracted into the film. About 4 per cent of the intensity of this refracted ray is internally reflected at the second face of the film and the remaining 96 per cent is transmitted by the film. The ray reflected at the second face of the film is finally refracted at the first face of the film and emerges from the film in a direction parallel to the ray reflected from the first face (Fig. 1.18(a)). These two rays will have approximately equal intensity but differ in optical path length. The former ray will suffer an optical path change of $\frac{1}{2}\lambda$, where λ is the wavelength of the light, due to the phase change of π on reflection from the first face of the film, which is the medium of higher refractive index. The latter ray will have an additional path length due to the additional path in the film and no phase change due to reflection at the second face. The two parallel rays will interfere. All rays incident on the film at the same horizontal level will interfere with the same phase, and path, difference between the two split rays. Interference by rays incident at different vertical film heights will have different phase differences. This phase difference will increase as the points of incidence of the rays moves down the film due to the increasing thickness of the film. We obtain bright horizontal bands due to constructive interference. This occurs when the phase difference between the split rays is a multiple of 2π. Dark horizontal bands, due to destructive interference, occur when the phase difference between the split rays is an odd multiple of π. These alternate interference bands of bright and dark light are best viewed against a black background in order to eliminate extraneous light.

When the vertical frame is withdrawn from the soap solution the initial angle of the 'wedge' will be relatively large (Fig. 1.18(a)). A large number of horizontal fringes will be visible. As the water drains away from the film, the angle of the wedge and the thickness of the film will decrease (Fig. 1.18(b)), and the number of interference fringes will decrease. Eventually only one fringe will be present and the film will begin to darken in the thinnest region at the top of the film. Finally after continuing to thin the film will become completely dark. This destructive interference results from a phase difference of approximately π, corresponding to a path difference of $\frac{1}{2}\lambda$, between the two split rays. This is due to the additional phase difference produced by reflection, at the first surface of the film, by the first split ray. The additional phase difference due to the second split ray entering the soap film is small compared with π.

The film eventually reaches an equilibrium thickness in which both faces of the film are parallel (Fig. 1.18(c)). In this state there is no variation in intensity over the surface of the film. This is called the *common black film*. It occurs, typically, at a thickness of 300Å (30nm). A further decrease in the film thickness, to another stable equilibrium state with a thickness of about 50Å, is often possible and is known as the *Newton black film*. This film is darker than the first black film, the common black film, as the second of the two split rays, that is refracted into the film, travels through a thinner soap film than in the case of the common black film. Consequently the phase difference between the two split rays of the Newton black film is closer to π than in the case of the common black film. Some films have only one equilibrium state while others have two or more equilibrium states.

The intensity of a light beam, I, reflected from a soap film of refractive index, μ, thickness, t, for an angle of refraction θ, is given by[32, 33]

$$I = I_0 \sin^2\left(\frac{2\pi\mu t}{\lambda}\cos\theta\right), \tag{1.28}$$

where I_0 is the maximum intensity produced by constructive interference and λ is the wavelength of the light. The angle of refraction, θ, is related to the angle of incidence of the light by Snell's law of refraction. Equation (1.28) can be used to calculate the thickness of the film, t, from a measurement of I and a knowledge of the order of interference.

A similar oscillatory behaviour in the variation of intensity applies to the interference resulting from the transmission of light through the film. In this case the additional phase factor of π, due to reflection at a medium of higher refractive index, does not occur. Thus a film with a thickness that is much smaller than the wavelength of light will give rise to a maximum in the intensity distribution and a minimum as occurs for the reflected light. The intensity of the light transmitted directly through the film is considerably greater than that due to the secondary ray, produced after two reflections in the film. Consequently when the two rays are out of phase the intensity is reduced slightly below the maximum intensity, which results from constructive interference. This will be discussed in greater detail in Chapter 2. The discussion has been concerned with a vertical film but with small modifications applies to films at any orientation to the horizontal.

In order to study the interference produced by reflection and transmission from a soap film it is advisable to keep the film in a controlled environment in order to prevent the film rupturing.

When a soap film finally ruptures it will break up into many small droplets. Most of the energy of the film is converted into kinetic energy of the droplets. This produces droplets with typical velocities of the order of 10^3 cm per sec.

1.9   Contact Angle, Dupré's Equation, and Neumann's Triangle

The surface of a liquid makes contact with a solid surface at a fixed angle. This angle, a, is the *contact angle*.[119, 120] It is the angle between the tangent planes to the liquid and the solid at any point along the line of contact (Fig. 1.19).

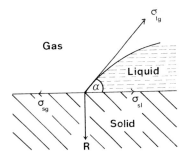

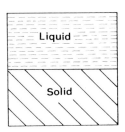

Fig. 1.19 Angle of contact, α, and the
surface tension forces for a
solid–liquid–gas system.

Fig. 1.20 A solid and liquid in contact.

In the example of a solid–liquid–gas system (Fig. 1.19), there will be three surface tension forces present at the line of contact of the three phases. The figure shows: the surface tension, σ_{lg}, between the liquid–gas phases directed at the contact angle, α, to the solid surface; the surface tension, σ_{sg}, between the solid–gas interface directed along the solid surface; the surface tension, σ_{sl}, due to the solid–liquid interface also directed along the surface of the solid but in the opposite direction to σ_{sg}. In order to maintain equilibrium the solid will produce a reaction, R, at the line of contact which is directed normally into the solid. Resolving these forces along the surface of the solid gives,

$$\sigma_{sl} + \sigma_{lg}\cos\alpha - \sigma_{sg} = 0. \qquad (1.29)$$

In the last century Dupré obtained another relationship between σ_{sl}, σ_{sg}, and σ_{lg} in terms of the work, ω_{sl}, required to separate a unit area of the liquid–solid interface. Consider a unit area of a liquid–solid surface (Fig. 1.20). Now, the energy of a unit area of interface plus the work done in separating the surfaces must be equal to the final energy of the unit areas of the solid surface and the liquid surface after they have been separated and are in contact with the gaseous phase. Thus using energy conservation Dupré's equation is obtained,

$$\sigma_{sl} + \omega_{sl} = \sigma_{sg} + \sigma_{lg}. \qquad (1.30)$$

On eliminating σ_{sl} and σ_{sg} by subtracting (1.29) from (1.30),

$$\omega_{sl} = \sigma_{lg}(1 + \cos\alpha). \qquad (1.31)$$

If a is zero the liquid is said to completely *wet* the solid and (1.31) gives

$$\omega_{sl} = 2\sigma_{lg}. \tag{1.32}$$

If $a = 180°$, the liquid does not wet the solid and (1.31) gives,

$$\omega_{sl} = 0. \tag{1.33}$$

However this is not experimentally realizable.

It is useful to introduce a wetting coefficient k defined by

$$k = \frac{\sigma_{sg} - \sigma_{sl}}{\sigma_{lg}}, \tag{1.34}$$

for the equilibrium of the liquid–solid–gas system. Putting $a = 0$ in Eq. (1.29), and substituting into (1.34) gives $k = +1$ if the solid is completely wetted and, from Eqs. (1.29) and (1.34), $k = -1$ if it is not wetted.

The discussion so far has concerned the equilibrium of a solid, a liquid, and a gas. If equilibrium of the three phases is not possible, there is no line of contact for the three phases. Equilibrium is impossible if no real value of a will satisfy Eq. (1.29), that is, for example, if

$$\sigma_{sg} > \sigma_{sl} + \sigma_{lg}. \tag{1.35}$$

The liquid will be pulled over the whole area by the resultant surface tension force, in the surface of the solid, and cover the solid surface. Alternatively, if

$$\sigma_{sl} > \sigma_{sg} + \sigma_{lg}, \tag{1.36}$$

the liquid–solid surface is displaced by the gas–solid surface. The liquid will remain on the surface as a drop without wetting the surface. The coefficient k can incorporate these *non-equilibrium* situations by defining $k \geqslant +1$ as complete wetting, $k \leqslant -1$ as not wetting.

When the solid is replaced by a liquid the results derived above remain valid provided that the liquids are immiscible. For a liquid in contact with itself $a = 0$, and from Eq. (1.31) $\omega_{ll} = 2\sigma_{lg}$. The work required to separate a unit area of two portions of the same liquid is $2\sigma_{lg}$. If $a = 0$ for a solid–liquid interface, from (1.32),

$$\omega_{sl} = 2\sigma_{lg} = \omega_{ll}. \tag{1.37}$$

This indicates that the attraction between the solid and the liquid is the same as the attraction between two parts of the same liquid. These results have not taken into account the frictional forces that are present when the liquid is on the point of advancing or receding over a solid surface. These forces can be taken into consideration by generalizing the arguments presented above.

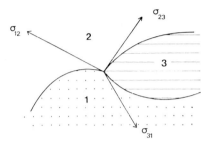

Fig. 1.21   Three fluids in equilibrium.

When three fluids, 1, 2, and 3 are in equilibrium the surface tension forces σ_{12}, σ_{23}, and σ_{31} at the line of contact of all three fluids must be in equilibrium (Fig. (1.21)). Thus for any small element of length, δl, along the line of contact of the three fluids,

$$\sigma_{12} + \sigma_{23} + \sigma_{31} = 0. \tag{1.38}$$

This 'triangular' equilibrium condition for the three vectors is known as *Neumann's triangle*. If a lens of liquid is sandwiched between two different fluids, the equilibrium of the fluids will appear as shown in Fig. 1.22 and the

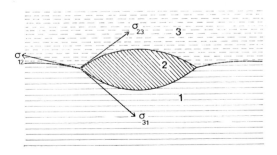

Fig. 1.22   A lens of fluid in equilibrium with two other fluids.

surface tension forces along the circle of contact of all three fluids will satisfy Eq. (1.38).

The general ground work required for the understanding of the subsequent chapters is now complete. In the following chapters we shall examine such subjects as the draining processes in soap films, the analogue solutions to minimum area problems, and the shapes of fluid surfaces.

2 DRAINING AND THINNING OF SOAP FILMS

2.1 Introduction

A freshly formed soap film contained by a frame will, typically, have a thickness of the order of a micron (10^3 nm, 10^4 Å), but it may exceed this value by as much as a factor of a hundred. Once formed the film will commence to drain. Mechanisms such as convection, evaporation, and suction produced by pressure gradients, will cause water to drain out of the film and result in the thinning of the film. These mechanisms can be divided into two main groups, static mechanisms and dynamic mechanisms. The static mechanisms are those in which the position of the surface of the film remains fixed, whilst the dynamic mechanisms produce movement of the surface. Evaporation is an example of a static process, and convection in the film is an example of a dynamic process as it produces the movement of the elements of the surface area of the film. If the film does not rupture these draining processes will continue until the thickness of the film has reached an equilibrium value. This is typically in the range 50 Å to 300 Å, which is the regime of destructive interference and blackening of the film when viewed by reflected light. The relative importance of the mechanisms responsible for draining will depend on the chemical composition of the film and such macroscopic quantities as the surface rigidity, surface viscosity, and environmental conditions such as temperature and pressure.

A freshly formed surface of a soap film reaches its equilibrium shape in the order of seconds. However the thickness of the film reaches its equilibrium value in a time that is orders of magnitude greater than that for the surface. For the fastest draining films, of low surface viscosity, this will be minutes whilst films of high surface viscosity will take hours to drain to their equilibrium thickness.

The property of the surface configuration to reach its equilibrium value, with minimum area, in seconds is used to obtain the solution of mathematical minimization problems in the following two chapters, and in Chapter 6 it is applied to the solution of Laplace's and Poisson's differential equation.

From the formation of the film until an equilibrium black film is obtained, the thickness can be measured by methods based on the interference of light. The interference produced by white and monochromatic light can conveniently be used to monitor the thickness of the film at any point on the surface during the thinning process. The order of the interference for monochromatic fringes gives an approximate guide to the thickness. The intensity of the light, given by Eq. (1.28), enables an accurate determination of the film thickness, t, to be made as

$$t = \frac{\lambda}{2\pi\mu \cos\theta} \sin^{-1}\left(\frac{I}{I_0}\right)^{1/2}, \qquad (2.1)$$

where λ is the wavelength, μ the refractive index of the film, I_0 the intensity for constructive interference, I the intensity of the reflected beam, and θ the angle of refraction. A knowledge of n, the order of interference, permits determination of t in (2.1).

If white light is used the interference pattern produced by the cumulative effect of interference, from each wavelength constituting white light, will give rise to colours in the film. The colour of each region of the film will be determined by its thickness. Consequently the thickness of the film at any point can be determined from the colour of the film. The variation of thickness over the whole film, at any time, can be mapped from the colours in the film.

This variation in thickness, over the surface of the film, as a function of time enables information concerning the draining and thinning mechanisms to be deduced. We shall now analyse the interference phenomena in some detail.

2.2   Interference Phenomena Produced by Soap Films

Figure 2.1 shows monochromatic light of wavelength, λ, incident at an angle i on a soap film of thickness, t. Some of the light will be reflected at the first face labelled B, and some will be transmitted by the film. These rays are indicated by $B_1 C_1$ and $B_1 D_1$ in Fig. 2.1. The transmitted light will be refracted on entering the film, the angle of refraction being θ. This ray will be partially reflected at the second face, D, at D_1 and finally emerge from the film at B_2 on the first face as ray $B_2 C_2$. The interference pattern will be largely determined by the two parallel rays $B_1 C_1$ and $B_2 C_2$ emerging from surface B. These rays have approximately equal intensity. We can neglect the variation in thickness of the film over the distance $B_1 B_2$. In order to obtain the optical path difference between these two rays it is convenient (Fig. 2.1) to construct

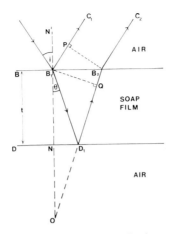

Fig. 2.1   Interference produced by reflection at a soap film.

perpendiculars from B_1 to D_1B_2 meeting D_1B_2 at Q, and from B_2 to B_1C_1 meeting B_1C_1 at P. In addition, the normal to the surface at B_1, B_1N, is extended to meet the extension of B_2D_1 at O. We shall assume that the refractive index of the air is unity. The optical path difference between rays B_1C_1 and B_2C_2, for a soap film of refractive index μ, is

$$\mu(B_1D_1 + D_1B_2) - B_1P. \qquad (2.2)$$

As $B_1D_1 = OD_1$, this is

$$\mu(OD_1 + D_1B_2) - B_1P, \qquad (2.3)$$

or

$$\mu(OQ + QB_2) - B_1P. \qquad (2.4)$$

Now in triangle QB_1B_2, as $Q\hat{B}_2B_1 = D_1\hat{B}_1B_2 = 90° - \theta$,

$$Q\hat{B}_1B_2 = \theta, \qquad (2.5)$$

hence

$$QB_2 = B_1B_2 \sin\theta. \tag{2.6}$$

In triangle B_1B_2P, as $B_1\hat{B}_2P = N'\hat{B}_1P$,

$$B_1\hat{B}_2P = i, \tag{2.7}$$

hence

$$B_1P = B_1B_2 \sin i. \tag{2.8}$$

Thus from (2.6) and (2.8),

$$\mu = \frac{\sin i}{\sin\theta} = \frac{B_1P}{QB_2}. \tag{2.9}$$

Substituting (2.9) into (2.4) the optical path difference becomes

$$\mu \, OQ. \tag{2.10}$$

In triangle B_1OQ,

$$OQ = OB_1 \cos B_1\hat{O}Q, \tag{2.11}$$

as $N\hat{B}_1D_1 = N\hat{O}D_1$,

$$= OB_1 \cos\theta, \tag{2.12}$$

as $OB_1 = 2NB_1$,

$$= 2t \cos\theta. \tag{2.13}$$

Thus the optical path difference, result (2.10), is

$$2\mu t \cos\theta. \tag{2.14}$$

In addition to the optical path difference due to the difference in path lengths between the two rays there will be an additional path difference of $\frac{1}{2}\lambda$ due to the additional phase difference of π that occurs at the air–film interface whenever an incident ray is reflected by a medium of higher refractive index than the initial medium. Thus the effective path difference between the two rays is

$$2\mu t \cos \theta + \tfrac{1}{2}\lambda. \qquad (2.15)$$

Consequently if

$$2\mu t \cos \theta + \tfrac{1}{2}\lambda = n\lambda, \qquad (2.16)$$

where n is an integer, the two rays will interfere constructively and give an intensity maximum. However, if

$$2\mu t \cos \theta + \tfrac{1}{2}\lambda = (n+\tfrac{1}{2})\lambda, \qquad (2.17)$$

there will be destructive interference resulting in zero intensity.

As the amplitudes, A, of the two waves are assumed to be equal, the resultant amplitude and phase of the reflected wave will be given by, A_r, where

$$A_r = A + Ae^{i\delta}, \qquad (2.18)$$

and the phase difference, δ, is given from (2.15) by,

$$\delta = \frac{2\pi}{\lambda}(2\mu t \cos \theta + \tfrac{1}{2}\lambda). \qquad (2.19)$$

The total reflected intensity, $I_r = A_r A_r{}^*$, is

$$I_r = (A + Ae^{i\delta})(A + Ae^{-i\delta}), \qquad (2.20)$$

$$= A^2(2 + e^{i\delta} + e^{-i\delta}), \qquad (2.21)$$

$$= 2A^2(1 + \cos \delta), \qquad (2.22)$$

as $\cos \delta = \tfrac{1}{2}(e^{i\delta} + e^{-i\delta})$.
Using the half angle formula, $(1 + \cos \delta) = 2\cos^2 \tfrac{1}{2}\delta$,

$$I_r = 4A^2 \cos^2 \tfrac{1}{2}\delta. \qquad (2.23)$$

From (2.19) this gives

$$I_r = 4A^2 \cos^2\left(\frac{2\pi}{\lambda}\mu t \cos\theta + \tfrac{1}{2}\pi\right).$$
(2.24)

This can be re-expressed in terms of the incident intensity, I_i, and the reflectivity of the soap film, $\mathcal{R}$. $\mathcal{R}$ is the fraction of the incident intensity reflected by the soap film and depends on the angle of incidence and the refractive index of the film. It can be obtained from classical electromagnetic theory and was first derived by Augustin Jean Fresnel[32] in 1823. Expressing (2.24) in terms of $\mathcal{R}$ and I_i,

$$I_r = 4I_i\mathcal{R} \cos^2\left(\frac{2\pi}{\lambda}\mu t \cos\theta + \tfrac{1}{2}\pi\right),$$
(2.25)

$$= 4I_i\mathcal{R} \sin^2\left(\frac{2\pi}{\lambda}\mu t \cos\theta\right).$$
(2.26)

Thus if θ is kept constant and t is varied there will be a fringe system in which each fringe corresponds to a region of constant t. This occurs in the case of a freshly formed vertical soap film where the thickness, t, of the film varies with the vertical height. The cross-section of the film is wedge-shaped as shown in Fig. 1.18(a). Plate 2.1 shows the horizontal system of fringes produced by a vertical soap film, illuminated by monochromatic light. If the upper portion of the film has a thickness that is negligible compared with the wavelength, λ, it will be black due to destructive interference caused by the additional phase difference of π due to reflection at the front face of the film. It is seen from Eq. (2.26) that $I_r = 0$ when $t \ll \lambda$. A horizontal soap film of constant thickness will have uniform intensity, for a fixed θ, given by (2.26). When the thickness alters, the intensity will also change, passing through maxima and minima given by Eq. (2.26).

When the film is illuminated by white light, coloured fringes will be observed. White light is composed of all the colours of the visible spectrum. Consequently the intensity of the emerging beam is the sum of the contributions from each wavelength, each wavelength will have an emerging intensity given by (2.26). The interference produced by an angle of refraction θ, for a film of thickness, t, much less than the wavelength of visible light, λ, will produce zero intensity and so the film will appear black. Each wavelength constituent of white light will interfere destructively as described earlier. So the resultant intensity due to the sum of all wavelengths will also be zero. As the thickness of the film is increased the shorter wavelengths, at the violet end of the spectrum, will be the first to interfere constructively and give rise to a bright interference fringe. This will be followed by the colours blue, green, etc. through the spectrum in the order of increasing wavelength. The separation of bright fringes by dark fringes, that occurs with monochromatic light, will be absent. The colour and intensity at any point on the film will be determined by the sum of contributions, given by Eq. (2.24), from all the visible wavelengths. The colour of the film ranges from black, for film thickness in the range 50–300 Å, to silvery white as the thickness of the film increases from 300 to 1200 Å. This is shown in Plates 2.2 and 2.3 at the centre of the book. It shows up more clearly in Plate 2.3 where the film has been allowed to drain and thin so that the variation in thickness of the film with height is reduced. This increases the height of film associated with any interference colour.

As the thickness of the film increases the interference produced by white light gives rise to a pale yellow section of film. This occurs in the region where the violet light approaches its minimum intensity. Other wavelengths will interfere constructively to give the resultant yellow colour. The thickness of the film is about 1200 Å, Plates 2.3 and 2.2. As the film thickness increases, each wavelength will give rise to an intensity contribution that will pass through successive stages of constructive and destructive interference. For any thickness it is necessary to add up the contributions to the intensity from all wavelengths. Plates 2.2 and 2.3 show the successive colours that occur in a draining vertical soap film in which the thickness increases with decreasing vertical height. Thus it is possible to observe the colours associated with a range of soap film thicknesses. Plates 2.2 and 2.3 give the order of the interference, n, on the left hand side. That is; $n = 0 \rightarrow 1$ (Eq. 2.16) for first order; $n = 1 \rightarrow 2$ for second order; etc. On the right hand side of the Plates is an indication of the thickness of the film for the different colours. Beyond the sixth order the wavelengths of different orders overlap appreciably.

Table 2.1 is taken from Lawrence's book[8] *Soap Films*. It contains a descriptive indication of the colours produced by the interference of white light. The white light is incident normally on a vertical soap film of refractive index $\mu = 1.41$. The Table also contains the thickness of soap film associated with each colour of film. For an incident beam which has an angle of refraction θ, an additional factor of $\cos \theta$ is required, Eq. (2.26), in order to obtain the thickness of the film from its colour in Table 2.1. Thick soap films will produce overlap of different orders of interference for the different spectral colours. The film will consequently appear increasingly white as the correlation between the interference produced by different colours decreases to zero.

TABLE 2.1

Interference colours produced by white light and the corresponding
thicknesses of the soap film.

First Order	Thickness in A	Fourth Order	Thickness in A
Black	$\begin{cases} 60 \\ 120 \end{cases}$	Grass Green	5970
Silvery White		Green	6340
	—	Yellow Green	6820
Amber	—	Carmine	7460
Magenta	2010		
		Fifth Order	
Second Order		Green	7900
Violet	2160	Green	8420
Blue	2500	Pink	8930
Green	2900	Pink	9450
Yellow	3220		
Orange	3480	*Sixth Order*	
Crimson	3710	Green	10000
		Green	10440
Third Order		Pink	11000
Purple	3960	Pink	11500
Blue	4100		
Blue	4280	*Seventh Order*	
Emerald Green	4660	Green	12100
Yellow Green	5020	Green	12650
Carmine	5420	Pink	13150
Bluish Red	5780	Pink	13700
		Eighth Order	
		Green	14200
		Pink	15000

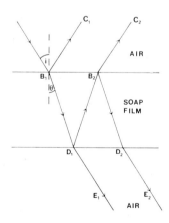

Fig. 2.2   Interference produced by transmission through a soap film.

In examining the interference due to white light the angle of incidence, i, and hence the angle of refraction, θ, have been kept constant. A variation in these angles will produce a variation in the phase difference, δ, Eq. (2.19), and hence a change in the interference pattern. For example a horizontal film with constant thickness, t, will produce a striking variation in colour when viewed at different angles of incidence. The colour of the film, when viewed by reflection, will depend on the phase difference only between the two light ray paths. Consequently the colour viewed at equal angles of incidence will be identical to that viewed by a vertical film providing the phase difference, $2\mu t \cos \theta$, is the same in both cases.

The same analysis can be applied to the interference pattern produced by transmitted light. Here the interference is due primarily to the two transmitted rays, indicated in Fig. 2.2, which emerge at D_1 and D_2. However the two transmitted rays have significantly different amplitudes. The intensity of the primary beam, that is transmitted without reflection, is at least an order of magnitude greater than the secondary beam that undergoes two reflections in the film. Consequently the intensity due to destructive interference does not result in zero intensity. A similar argument to that for reflected rays shows that the path difference between the two emerging transmitted rays is

$$2\mu t \cos \theta. \tag{2.27}$$

There is no additional contribution of $\frac{1}{2}\lambda$ due to reflection at an air–film interface. The only reflections occur in the soap film medium at film–air interfaces. These reflections produce zero phase difference. The intensity of the transmitted beam, I_t, can be obtained from the result for reflected rays, (2.26), as the sum of I_r and I_t must equal the total incident intensity, I_i, if no absorption of light by the soap film takes place. Thus

$$I_i = I_r + I_t. \tag{2.28}$$

Hence from (2.26) and (2.28),

$$I_t = I_i\left(1 - 4\mathscr{R} \sin^2\left(\frac{2\pi}{\lambda}\mu t \cos \theta\right)\right). \tag{2.29}$$

When $t = 0$ the intensity will be a maximum for all wavelengths. All wavelengths will interfere constructively. The two interference patterns produced by the reflected and transmitted light are complementary. That is they must add up to the incident white light beam (Eq. (2.28)).

This theory is accurate for most practical experimental purposes. However approximations have been made which require further discussion. In examining the interference produced by reflection and transmission only two waves, the most important, have been taken into account. There will be an infinity of waves producing interference by reflection (Fig. 2.3) C_1, C_2, C_3 ... etc. and similarly for transmitted waves E_1, E_2, E_3, ... etc.

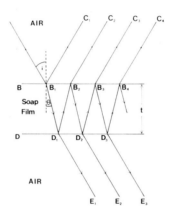

Fig. 2.2 Interference produced by transmission and reflection by multiple waves.

In order to examine the importance of the third wave and higher order waves it is necessary to introduce reflection factors r and r' and transmission factors τ and τ'. r is the ratio of the reflected amplitude to the incident amplitude of the wave in air when it is reflected by the soap film surface. r' is the reflection factor for an incident wave from the soap film medium. τ is the ratio of the amplitude of the transmitted wave to that of the incident wave in propagating from air to the soap film and τ' is the factor for transmission from the soap film to the air. If A_i is the amplitude of the incident wave and δ is the phase difference between successive rays emerging from the film, the total amplitude reflected from the soap film will be obtained by adding the contributions from each reflected ray. This is

$$rA_i + \tau\tau'r'A_ie^{i\delta} + \tau\tau'r'^3A_ie^{2i\delta} + \ldots \tau\tau'r'^{(2p-3)}A_ie^{i(p-1)\delta} + \ldots, \qquad (2.30)$$

where $p = 2, 3, 4 \ldots$ gives the general term following rA_i. Similarly for transmitted waves $E_1, E_2, E_3, \ldots$ the total transmitted amplitude is determined by

$$\tau\tau'A_i + \tau\tau'r'^2 A_i e^{i\delta} + \tau\tau'r'^4 A_i e^{2i\delta} + \ldots \tau\tau'r'^{2(p-1)} A_i e^{i(p-1)\delta} + \ldots \tag{2.31}$$

where $p = 1, 2, \ldots$.

The reflection coefficients r and r' are related by

$$r = -r'. \tag{2.32}$$

The negative sign produces a phase change of π in waves reflected at the first face of the soap film, as discussed earlier. This phase change always occurs when a ray is reflected in a medium of lower refractive index than that at which the reflection takes place. The reflectivity is given by

$$\mathscr{R} = r^2 = r'^2. \tag{2.33}$$

The transmissivity, $\mathscr{T}$, the fraction of the incident intensity transmitted by a surface of the soap film, is given by

$$\mathscr{T} = \tau\tau'. \tag{2.34}$$

As the energy of the light wave is conserved we must have,

$$\mathscr{R} + \mathscr{T} = 1. \tag{2.35}$$

In examining the interference produced by reflection we have assumed that only the first two contributions in (2.30) are important and that they have equal amplitudes. This is true if $r' \ll 1$ and $\tau\tau' \sim 1$, and is usually justified experimentally. Likewise in (2.31), if $r' \ll 1$ it is only necessary to consider the first two contributions. The magnitudes of the first two terms in the reflected amplitude, (2.30), are equal. However for transmitted rays, (2.31), the magnitude of the two contributions will not be equal, the first contribution being at least an order of magnitude greater than the second.

Corrections[32] to (2.26) and (2.29) can be made to take account of the additional rays $C_3, C_4, C_5 \ldots$ and $E_3, E_4, E_5 \ldots$ using the transmission and reflection factors that occur in (2.30) and (2.31).

2.3 Thinning by Static Mechanisms

The static mechanisms that produce the thinning of a soap film are those in which the position of the surface remains fixed, or in which the movement can be neglected. There are three basic static processes, stretching, viscous flow, and evaporation.

The typical soap film consists of two surfaces of amphipathic solute ions plus a distribution of solute ions in the water between the surfaces (Fig. 2.4(a)). It will be recalled that these amphipathic ions consist of two dissimilar parts. One part is hydrophilic with an affinity for water and tends to become surrounded by water molecules. This is the polar carboxyl 'head'. The other part

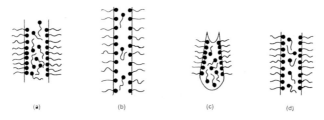

Fig. 2.4   Distribution of soap ions in: (a) original soap film; (b) a stretched film; (c) with viscous flow; (d) a film after evaporation has occurred.

of the amphipathic ion is hydrophobic with an aversion for water molecules and tends to avoid the presence of water molecules. This is the hydrocarbon 'tail' of the ion. Consequently there is a tendency for the carboxyl 'heads' to lie in the surface of the soap film and for the hydrocarbon 'tails' to point out of the surface. This, together with the metal ions of the amphipathic molecules, forms the intralamellar solution. When the soap film is stretched the surface density of the amphipathic ions will initially decrease as the surface has increased. However some of the intralamellar amphipathic ions will move to the surface in order to increase the surface density of soap ions. This will result in a decrease in the bulk density of soap ions (Fig. 2.4(b)). The number of amphipathic ions in the surface and the bulk will be determined by the requirements for thermodynamic equilibrium.

Viscous flow is illustrated in Fig. 2.4(c). The surface area of the film remains constant and its surfaces are rigid with respect to movement in their plane. The liquid between the surfaces can flow by simple laminar flow, as shown in Fig. 2.4(c). The film becomes wedge-shaped as the intralamellar fluid flows down under gravity. The assumption of a rigid surface is a good initial approximation, however experiment shows that it is more like a plastic surface.

Evaporation will thin the soap film by removing the water from the soap film. The decrease in the amount of water present will reduce the surface tension, by increasing the surfactant molecules, and cause the film to stretch (Fig. 2.4(d)). This can be easily observed by passing dry air from a jet over the surface of the soap film.

Variations in the interference colours, due to white light interference, will be produced by the thinning of the film. All these processes can be studied using the interference technique to obtain accurate measurements of the changes in thickness of the film.

2.4   Thinning by Dynamic Mechanisms

Josiah Willard Gibbs[52] was a careful observer of the dynamic mechanisms causing the thinning and draining of films. He observed that there were strongly turbulent regions near the borders of a soap film which are composed of thinner film than the central film (Plate 2.6). These thinner regions move under the action of gravity to higher levels with the same thickness. For a vertical soap film the higher regions of the soap film will be thinner than the lower regions as the cross-section of the film is wedge-shaped. These thinner regions, near the border, move upwards to levels of equal thickness where they remain in relatively stable equilibrium. This process is known as *gravity convection*.

At the border of a soap film, where it makes contact with a wire frame, the curvature of the film is concave outwards (Fig. 2.5(a)). This curvature corresponds to a negative pressure difference according to the Laplace–Young equation, section 1.5. The fluid in the border region is at a lower pressure than the surrounding atmospheric pressure. This border is known as the *Gibbs ring* or *Plateau border*. A similar region occurs at the intersection of three soap films (Fig. 2.5(b)). This negative pressure region is evident as the vertical height of the border is above the level of the fluid in the bath so that the hydrostatic pressure at the border is less than that in the surface of the bulk fluid.

What is the origin of these thinner regions near the border of the soap film? It might appear that the negative pressure difference in the Plateau border 'sucks' in liquid from the adjacent soap film. However if we examine the film in greater detail we find that this is not possible. The rate of flow between the two surfaces of the soap film is slow as there is no pressure gradient along the flat portion of the film with zero curvature. So liquid can only be drawn from the small region adjacent to the Plateau border. Once this region has been thinned it would act as a constriction preventing any flow of liquid.

Fig. 2.5   Plateau border or Gibbs ring: (a) at the boundary of a soap film and a rigid wire; (b) at the junction of three soap films.

Fig. 2.6   The process of marginal regeneration.

Careful observation by Professor Karol J. Mysels[34] and collaborators has shown that the thicker film is drawn *bodily* into the border by the negative excess pressure, ΔP, and the thinner film is pulled out of the border. This is illustrated in Fig. 2.6. If two films of different thickness are in contact with the border region, the thicker film will have the greater force drawing it into the border. This is illustrated in the left hand side of Fig. 2.6. The thinner film is drawn out of the right hand of the figure, the resistance to the motion is provided by the viscous forces. This draining mechanism occurs at adjacent regions of the border, often at regularly spaced sections. Mysels has called this process *marginal regeneration*. It is shown in a horizontal film in Plate 2.4, and in Plate 2.6.

2.5   Classification of Soap Films

Professor Mysels has distinguished soap films with three extreme types of draining behaviour. They are the *rigid film*, the *simple mobile film*, and the *irregular mobile film*. As an example of the rigid film he used sodium lauryl sulphate below the c.m.c. of the pure material (0.15 per cent) and containing lauryl alcohol (8 per cent of dry lauryl sulphate). For an example of the simple mobile film he used pure sodium lauryl sulphate at concentrations from about the c.m.c. to 2.3 per cent by weight. Irregular mobile films were produced with concentrated solutions (6–8 per cent) of sodium lauryl sulphate.

These different characteristic films can often be produced with impure surfactants; washing-up liquids. The impure rigid film can often be prepared with the addition of a quantity of glycerine.

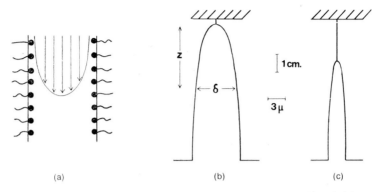

Fig. 2.7   The cross-section of a typical *rigid* draining soap film with a height of 8 cm and a cross-section of approximately 3 microns: (a) molecular profile of the flow; (b) cross-section of a soap film after 15 minutes; (c) cross-section of a soap film after 60 minutes.

The rigid film has surfaces that are rigid or plastic in nature and offer a high resistance to any motion. The rigidity of the surface can be gauged qualitatively by flicking a matchstick on the surface of the soap solution. It will rapidly come to rest showing that the surface has a high resistance to motion. When a vertical film is formed it does not show any visible turbulence in the surface and the film is relatively thick and slow draining. There may be no interference fringes initially. However after draining for a period the film will thin and interference fringes will appear. They will be closely spaced, showing that the film is relatively thick, and jagged with a horizontal trend (Plate 2.5). This indicates that the thickness of the film varies along a horizontal level due to the high viscosity and rigidity of the surface. As draining continues and the film thins the number of fringes, with a jagged appearance and with a horizontal trend, will decrease. A black film will eventually begin to form along the border and grow in area.

A rigid soap film created in an enclosure is prevented from thinning by evaporation. If it is allowed to drain in a saturated, vibrationless, clean, environment the film will not rupture. In such circumstances the drainage will be by viscous flow alone. The fluid will flow between the rigid, or plastic surfaces of the soap film (Fig. 2.7(a)). The profile of the film will be parabolic as indicated in Fig. 2.7(b). The parabolic cross-section is joined at its highest point to avertical black film (Fig. 2.7(b)). The black film will have a constant thickness which is negligible compared with the parabolic part of the soap film under-going viscous flow.

The relation between the thickness of the parabolic section of film, t, the vertical distance, z, measured downwards from the top of the parabolic film (Fig. 2.7(b)), and the time, τ, measured from the instant of creation of the film, is[34]

$$t^2 = \left(\frac{4\eta}{\rho g}\right)\left(\frac{z}{\tau}\right), \qquad (2.36)$$

where ρ is the density of the soap solution and η is the viscosity of the film. The bulk values are usually taken as approximations for ρ and η. Figure 2.7(b) gives a typical cross-section for a rigid film 15 minutes after its creation. Figure 2.7(c) gives the cross-section after 60 minutes. The vertical line above the parabolic cross-section indicates the thin black film.

The simple mobile film drains much more rapidly than the rigid film and is the commonest type of soap film. Horizontal interference bands are observed immediately after creation of a vertical film (Plates 2.2 and 2.3). This indicates that the film has constant thickness at any horizontal level. The film shows rapid turbulent motion at the borders due to marginal regeneration and gravitational convection (Plate 2.6). Viscous flow effects can be neglected in the contribution to the draining of the simple mobile film. The variation in the thickness of a typical simple mobile film with time is shown in Fig. 2.8. The draining and thinning process is usually an order of magnitude faster than that in a rigid film. Rapid thinning occurs along the horizontal and vertical borders. At the top of the film a black film appears, and is represented by the vertical line in the film profile (Fig. 2.8). The primary cause of thinning is the vertical border. If we examine the draining of a vertical cylindrical film, with no vertical border, the draining and thinning process will be appreciably reduced.

The boundary between the black and silver region is horizontal except near the vertical border. This sudden change in colour from black to silver at a horizontal level indicates that the thickness of the film has altered rapidly. This region is indicated in Fig. 2.8 by the horizontal level at which the thin black film meets the draining film. This is an important characteristic that distinguishes the simple mobile film from the irregular mobile film. The rapid variation in the thickness of the film at the black–silver border occurs when the thickness of the film becomes comparable to the range of the van der Waals interaction between the water molecules in the film.

The irregular mobile film initially drains in a similar manner to the simple mobile film. The differences appear after the formation of the black film. The boundary between the black region and the coloured region becomes unstable. Streamers of black film encroach into the coloured region of the film and end in coloured islands of film (Plate 2.7). This phenomenon was called *critical fall* by Sir James Dewar.[8, 53] The explanation for this behaviour is given in the section 2.7. The black film will grow and eventually spread throughout the whole film.

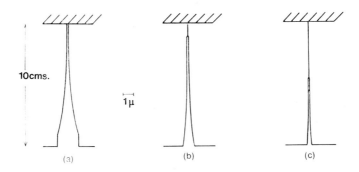

Fig. 2.8   The cross-section of a typical simple mobile film with a height of 10 cm and cross-section of approximately 1 micron: (a) after 40 seconds; (b) after 120 seconds; (c) after 240 seconds.

These are the three main categories of soap film behaviour. Soap films need not fall into one of these three but can have a behaviour of an intermediate character. For example a soap film could have a behaviour typical of a simple mobile film with a small amount of critical fall.

2.6   The Black Film

In a controlled environment a soap film will thin until its thickness becomes appreciably less than the wavelength of light so that it appears black when viewed by reflected light. Commonly it is found that the film reaches a stable thickness, in thermodynamic equilibrium, of about 300 Å. This is known as the *common black film*. If a small amount of evaporation is allowed to take place the film will often thin to a new equilibrium thickness of about 50 Å. This is the *Newton black film*. The thickness of these two equilibrium states can be obtained by measuring the intensity of the reflected light by using photoelectric detectors. The application of Eq. (2.26) will enable the thickness of the film, t, to be determined.

Although the two black films are common to many soap films it is possible to have only one stable black region. The stability of the black film will depend on the molecular forces present in the film. This will differ for different surfactant molecules and hence for different soaps. It is also possible to obtain soap films that behave more like solid flakes. They are capable of forming stable black films by stratification. Each layer consists of a basic unit which is a bimolecular leaflet, similar in structure to that of the Newton black film, as indicated in Fig. 2.9. It consists of two surfactant layers of ions separated by a distance of about 30–35 Å. Using this basic unit it is possible to obtain stable black films with thicknesses which are multiples of the basic thickness. This stratification can extend into the silver region.

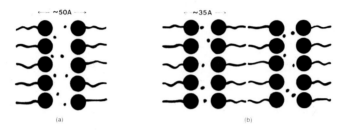

Fig. 2.9   The molecular structure of soap films: (a) the Newton black film; (b) a stratified film consisting of two 'bimolecular leaflets'.

The simple mobile film usually has a black–silver boundary due to the rapid increase in the thickness of the film. In concentrated soap solutions it is often found that the boundary is black–yellow followed by a silver region. This is shown diagrammatically in Fig. 2.10 and is easily explained. Initially the mobile film will form a black–silver region as shown in Fig. 2.10(a). As the area of the black film increases, excess liquid accumulates at the front of the expanding black film. This produces a region that is thicker than the previous silver region, as indicated in Fig. 2.10(b), which will produce the yellow interference region of the film. Lower down the film, where the excess fluid has not accumulated, is a thinner region which appears silver.

2.7   Critical Fall and the Irregular Mobile Film

The main feature of an irregular mobile film is the erratic behaviour of the black–silver region. Rivers of black film, with tributaries, appear to break through into the coloured film (Plate 2.7). These usually produce small areas of highly coloured film surrounded by black film. The general appearance of the film is reminiscent of peacock feathers. There is a considerable amount of convection due to the presence of the lighter black film in the region of the heavier coloured film. This produces rapid growth of the black region. This phenomenon of *critical fall* can be explained after careful observation of the film. It generally occurs in films formed from concentrated soap solution with more than 4 per cent of surfactant.

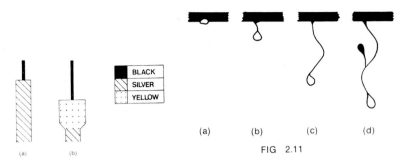

(a)   (b)   (c)   (d)

FIG 2.11

Fig. 2.10   (a) The cross-section of a black-silver film; (b) the cross-section of a black-yellow-silver film where fluid has accumulated in the yellow region.

Fig. 2.11   The stages in the production of critical fall: (a) the isolation of a region of silver film by the black film; (b) the resulting thickening of the silver film causes it to sink into the coloured film; (c) this causes the growth of a 'river' of black film; (d) 'tributaries' of black film form and rise to form 'lakes'.

The steps leading to critical fall are illustrated in Fig. 2.11. Initially, Fig. 2.11(a), a small region of the silver film becomes isolated and surrounded by black film. As the area of the black film surrounding it grows, it shrinks, becoming thicker and heavier. It falls, under gravity, into the main body of the coloured film without coalescing with it. It leaves a narrow 'river' of black film in its path (Figs 2.11(b) and (c)). The silver island of film continues to be surrounded by black film. Growth of the black film proceeds along the paths of the black film. This leads to 'tributaries' of the black film. These will rise as they are lighter than the surrounding coloured film. When the widths of the 'tributaries' are not great enough to cause an upward flow, black film accumulates in small 'lakes'. This accumulated black film may eventually cut its way through the coloured region and possibly reach the main body of black film. This whole cycle is repeated at many points along the black–silver border and leads to a rapid growth of the black film. The streamers of black film end around islands of heavier, highly coloured, film.

2.8   Potential Energy and Molecular Forces

A typical soap film is shown in Fig. 2.12. It contains anionic soap ions at the surface, water molecules, metal ions, and soap anions in the film. The hydrophilic 'heads' are commonly negatively charged but can be neutral or positively charged. The hydrophobic 'tails' are electrically neutral. The fluid in the soap film will have a net positive charge, for an anionic soap, as the film is electrically neutral. The attraction between the positive and negative charges will form a double layer of charge at each surface as indicated in Fig. 2.12.

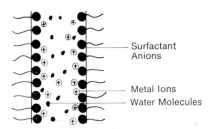

Fig. 2.12   Molecular structure of a typical soap film, containing anionic surfactant molecules plus metal ions and water molecules.

There are three important contributions to the molecular forces present in a soap film that have been investigated and explain some of the equilibrium properties of soap films, particularly the properties of the thicker equilibrium film, the common black film. These are the van der Waals attraction between molecules and ions, the electrostatic repulsion due to the double layers of charge at each surface, and the Born repulsion due to the hard cores of the molecules and ions. In addition there are other molecular forces in the soap film that are not fully understood. They are probably partly steric in nature, arising from the rigidity of the surfactant. The forces due to hydrogen bonding between the water molecules are not significant. They would become important if the equilibrium thickness of the soap film was less than 20 Å.

The van der Waals attraction between two water molecules is due to the induced dipole moment in the molecules and permanent dipole moments of the water molecules. The attractive potential energy between two molecules varies as the inverse sixth power of the distance between the molecules. The total van der Waals attractive potential energy of a soap film of thickness, t, due to the sum of the interactions between all the molecules in the film is given approximately by

$$V_a = -\frac{V_w}{t^2},$$ (2.37)

where V_w is a constant for a particular molecular system. The potential is thus an inverse square function of the thickness of the film. The negative sign indicates that the force is attractive.

The electrostatic repulsion is due to the overlap of the double layers of charge at each surface (Fig. 2.12). If the surfaces are sufficiently far apart there will be no appreciable overlap of the double layers. Each surface consists of a layer of negative surfactant ions and an associated positive charge, due to a diffuse layer of metal soap ions in the fluid, which are attracted to the negative surfactant ions. The total potential energy due to the repulsion between the double layers of charge at each surface of the film is given by

$$V_r = V_0 \exp(-\kappa t), \qquad (2.38)$$

where V_0 and κ are constants for the molecular system in thermodynamic equilibrium.

When the thickness of the film, t, is sufficiently small the Born repulsion potential, V_B, together with other less well understood contributions to the potential, U, become important. In the region of the common black film the V_B and U are negligible, but become significant in the region of the Newton black film.

The total potential, V, is the sum of U, V_B, V_r, and V_a so

$$V = U + V_B + V_r + V_a. \qquad (2.39)$$

From (2.37) and (2.38) this becomes

$$V = U + V_B + V_0 \exp(-\kappa t) - \frac{V_w}{t^2}. \qquad (2.40)$$

Figure 2.13 shows the variation of V with t to be expected for a typical film. For large t the van der Waals attractive potential is dominant. As t is reduced the double layers begin to overlap and the double layer repulsion becomes increasingly important and will give rise to the first minimum in the potential energy function. This stable thickness, where the potential energy has a minimum value, is the thickness associated with the common black film. A further reduction in t increases the importance of the double layer repulsion. However a stage will eventually be reached where the van der Waals attractive force again begins to dominate as it behaves as the inverse square of the thickness of the film and increases in magnitude with decreasing t. The potential energy thus decreases. A second minimum in the potential energy function is often obtained in the region where V_B and U are significant. For the further reduction in the thickness of the film the potential energy rapidly increases due to the strong repulsive forces due to V_B and U.

The two minima in the potential energy function, Eq. (2.40) and Fig. 2.13, are associated with the stable black films known as the common black film and the Newton black film. This behaviour is typical of a simple mobile film. However it is possible that a soap film could give rise to only one minimum in the potential energy function V. This will depend on the relative importance of the various contributions, in (2.40), to the total potential energy. In the case of a nonionic soap the double layer repulsion will be absent and will be replaced by a shorter range repulsion due to the interaction between the hydrophilic parts of the nonionic soap.

The van der Waals attractive forces between water molecules can explain the draining of a soap film in the region in which the van der Waals forces are dominant, and also the rapid change in thickness at the black–silver boundary of the interference pattern produced by reflected white light. Figure 2.14 represents the cross-section of such a soap film. A water molecule such as W in Fig. 2.14 will experience van der Waals attractions from all its neighbouring molecules. The forces on W due to those neighbours within a sphere, S_1, will tend to cancel each other. For example, the attraction due to molecule X (Fig. 2.14) will cancel that from a molecule diametrically opposite at X'. Outside the sphere S_1 there are molecules, such as Y and Z in the soap film, which do not have any diametrically opposed molecules as the soap film is wedge-shaped. The attractions due to such molecules will not cancel. They will give rise to a resultant force on W directed towards the thicker part of the film. Thus the narrower part of the film cross-section becomes thinner as fluid drains into the thicker part. This results in an abrupt variation in thickness of the soap film in the region where its thickness is comparable to the range of the van der Waals forces. It produces the sharp black–silver border produced by white light reflections from the film.

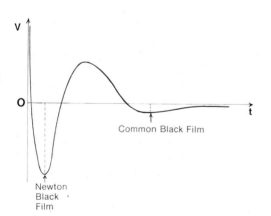

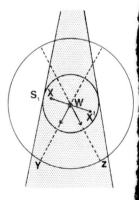

Fig. 2.13   T hetotal potential energy, V, of a soap film as a function of thickness, t.

Fig. 2.14   Drainage of a soap film due to Van der Waals forces.

1.1 (a)

1.1 (b)

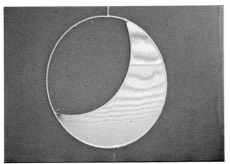

1.1 (c)

1.1 Planar minimum area surfaces bounded by a circle.
(a) A disc of soap film.
(b) The minimum surface formed by breaking the soap film inside the loop of thread.
(c) The minimum surface formed between the circle and a length of thread.

2.1 Interference fringes produced by a vertical soap film using monochromatic light.

2.2 Interference colours produced by a vertical soap film, shortly after formation, using white light.

2.1

2.2

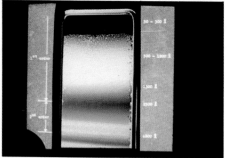

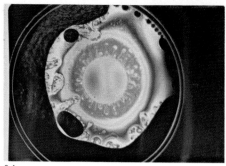

2.3

2.4

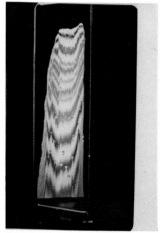

2.3 Interference colours produced by a vertical soap film, after draining for some time, using white light.

2.4 Marginal regeneration in a horizontal film.

2.5 A vertical rigid film.

2.6 A vertical simple mobile film draining along a border due to marginal regeneration and gravity convection.

2.7 A vertical irregular mobile film displaying critical fall.

2.5

2.6

2.7

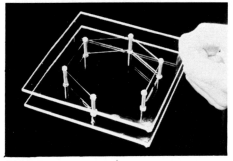

3.1 (a)

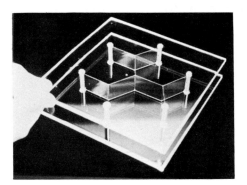

3.1 (b)

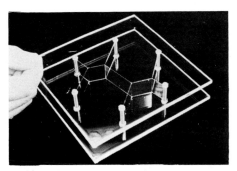

3.1 (c)

3.1　Minimum paths linking the vertices
of a hexagon.
(a) Absolute minimum path.
(b) Path with three-fold axis of
symmetry.
(c) Path with two-fold axis of
symmetry.

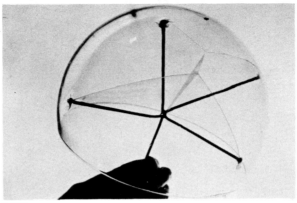

3.2 (a)

3.2 (b)

3.2 Minimum paths linking towns on
the surface of the earth.
(a) Four towns.
(b) Three towns.

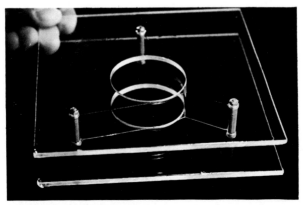

3.3 Minimum path linking three towns
and avoiding a circular constraint.

4.1

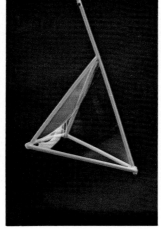

4.2

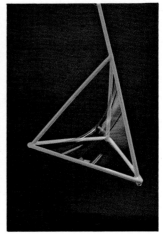

4.3

4.4

4.5

4.1 The minimum surface formed by a soap film contained by the edges of a tetrahedron.

4.2 The minimum surface formed by a soap film contained by five edges of the tetrahedron.

4.3 The soap film surface formed by four edges of the tetrahedron.

4.4 The soap film surface formed by the twelve edges of a cubic framework.

4.5 A minimum surface formed by breaking the minimum surface in Plate 4.4 at two sections of surface linked to two edges.

4.6 (a)

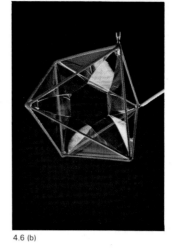

4.6 (b)

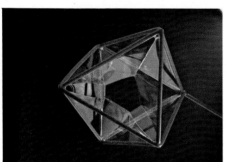

4.6 (c)

4.6   Minimum surfaces bounded by the
twelve edges of the octahedron.
(a) Central point
(b) Central 'hexagon'
(c) Central non-regular 'pentagon'
(d) Central 'square'
(e) Surface containing a
'quadrilateral'

4.6 (d)

4.6 (e)

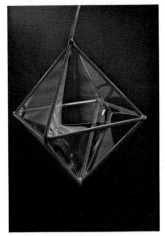

4.7 (a)

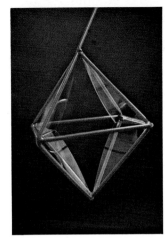

4.7 (b)

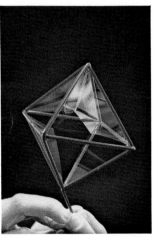

4.7 (c)

4.7 Some of the minimum surfaces bounded by a subset of edges of the octahedron.
(a) Asymmetric surface
(b) Internal 'rhombus'
(c) Surface containing a planar kite.

4.8
(a) The minimum surface formed inside a dodecahedral framework.
(b) The minimum surface formed by an icosahedral framework.

4.8 (a)

4.8 (b)

4.9 (a)

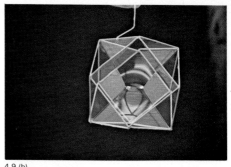

4.9 (b)

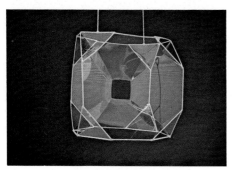

4.9 (c)

4.9 (d)

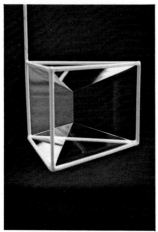

4.10

4.9   Some surfaces formed by Archimedean Frameworks.
   (a) The minimum surface inside a truncated
   tetrahedron
   (b) A minimum surface inside a cuboctahedron
   (c) The minimum surface inside a truncated cube
   (d) A minimum surface inside a truncated octahedron

4.10 The minimum surface, contained by a triangular prism
   framework, with a central vertical axis.

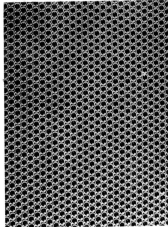

4.12 (a)

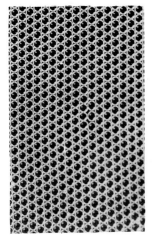

4.12 (b)

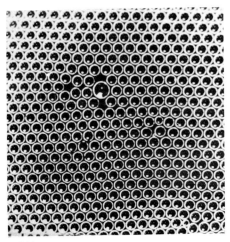

4.12 (c)

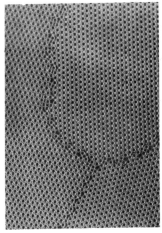

4.12 (d)

4.12 Bubble Rafts.
  (a) Perfect crystalline raft
  (b) Raft with a dislocation
  (c) Raft with a defect
  (d) Raft with grain boundaries

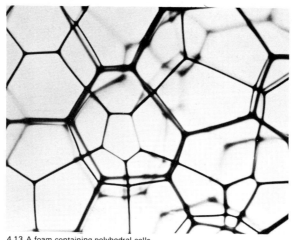

4.13 A foam containing polyhedral cells.

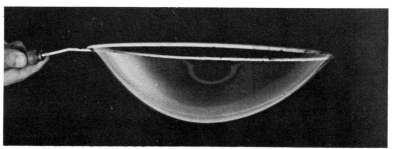

6.1 (a)

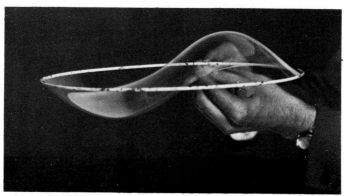

6.1 (b)

6.1 Normal modes of a circular
membrane.
(a) (0,1) mode
(b) (1,1) mode.

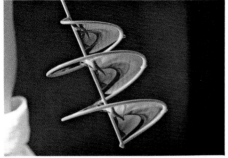

4.11

4.14 (a)

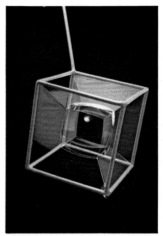

4.14 (b)

4.14 (c)

4.14 (d)

4.11 The minimum surface contained by a helix with a central axis.

4.14 Bubbles contained by frameworks.
(a) A bubble inside a tetrahedral framework
(b) A bubble inside a cubic framework
(c) A bubble inside an octahedral framework
(d) A bubble inside a triangular prism framework

4.14 (e)

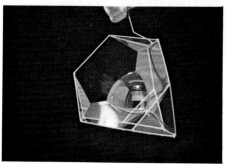

4.14 (f)

4.14 Bubbles contained by frameworks.

    (e) A bubble inside a dodecahedral framework
    (f) A bubble inside a truncated tetrahedron
    (g) A bubble inside a cuboctahedron
    (h) A bubble inside a truncated cube

4.15 (a) and (b)

    A 'spherical' bubble being bounced
    off a 'disc' of soap film.

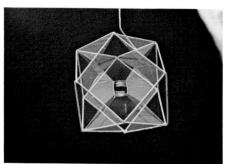

4.14 (g)

4.14 (h)

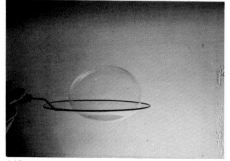

4.15 (a)

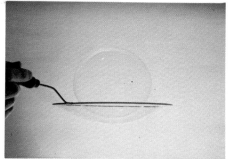

4.15 (b)

3 *THE MOTORWAY PROBLEM*

3.1 Free Energy of a Soap Film

A soap film contained by a fixed boundary, such as a soap film formed within a circular ring, can change its shape and area. If the area of the film is A and its film tension is σ_f, then the energy available to alter its shape at constant temperature and constant σ_f is

$$F = \sigma_f A. \tag{3.1}$$

F is known as the *free energy*. When the film is in equilibrium, or more correctly thermodynamic equilibrium, the free energy will be minimized. This is analogous to the minimization of the potential energy of a system of particles in classical mechanics. From Eq. (3.1) the area, A, of the film will have a minimum value when the film has reached equilibrium as σ_f remains constant. The film tension, σ_f, is a function of temperature only.

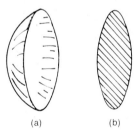

(a)   (b)

Fig. 3.1   Soap films contained by a circular ring; (a) a non-equilibrium soap film; (b) the equilibrium soap film having the shape of a disc.

It is possible to use the result concerning the minimization of the surface area of a soap film to solve some mathematical minimization problems. In order to do this we must construct a boundary with the property that the solution to the mathematical problem will be given by the area of the soap film, when it is minimized, at equilibrium. In the case of a soap film contained by a circular ring the equilibrium configuration gives the solution to the mathematical problem of determining the minimum area bounded by the ring. We know this to be the disc contained by the ring. Hence we expect the film to form a disc contained by the ring once it has come to rest. Indeed this is the case, as shown in Fig. 3.1 and Plate 1.1(a).

We shall examine, using soap films, the analogue solutions to some two dimensional mathematical problems. These are problems that are restricted to a surface.

3.2 The Motorway Problem

One of the mathematical results that we all learn early in life concerns the shortest path linking two points. This path is the straight line joining the two points. If one attempts to generalize this result by increasing the number of points the problem becomes increasingly more difficult. What, for example, is the shortest path joining three points, or four points, or more points? In fact the general problem of connecting an arbitrary number of points by the *shortest* path has not been solved analytically and has to be determined by computation.

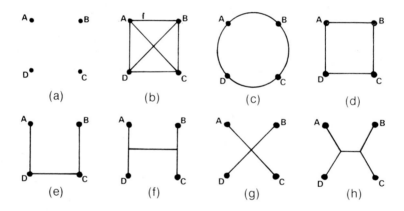

Fig. 3.2   Motorway configurations: (a) Four towns *A*, *B*, *C* and *D* at the corners of a square. (b) The roadway networkl inking any two towns by the shortest path. (c) The circular roadway. (d) The square roadway. (e) the U-shaped roadway. (f) The H-shaped roadway. (g) The X-shaped roadway. (h) The minimum length of roadway linking the four towns.

In order to gain some insight into these problems let us focus attention on one which might appear amenable to a simple solution. Consider the problem of linking four towns *A*, *B*, *C* and *D* by a road or motorway (Fig. 3.2(a)). For simplicity let the towns be situated at the corners of a square of unit length.

The network of roadways that will enable us to travel from any town to any other town by the shortest possible route is shown in Fig. 3.2(b). It consists of a network of roads along the edges of the square and along the two diagonals. Its total length is

$$4+2\sqrt{2} = 6.83.$$

This is obviously not the shortest length of road joining the four towns. A circular road passing through A, B, C and D (Fig. 3.2(c)), has a length of

$$2\pi(\tfrac{1}{2}\sqrt{2}) = 4.44,$$

which is considerably shorter than the 6.83 units of Fig. 3.2(b), and enables us to travel from any town to any other town. The total length of roadway can be further reduced by using the one mathematical result at our disposal; the result concerning the shortest distance between two points. Each circular arc, such as AB in Fig. 3.2(c), can be replaced by a straight line. This produces the square roadway configuration of Fig. 3.2(d) with a length of

$$4.00.$$

It is clear that a further reduction in length can be made by removing one of the sides of the square to give a U-shaped road, Fig. 3.2(e). This has a length of

$$3.00.$$

It is still possible to travel from any town to any other town. However the route from A to B will require us to travel 3 units. An H-shaped configuration, such as Fig. 3.2(f), has the same total length but reduces the journey from A to B to 2 units.

From the symmetry of the square we might be tempted to conjecture that the shortest road is the cross-roads system of Fig. 3.2(g). It has a length of

$$2\sqrt{2} = 2.83.$$

This is the shortest roadway encountered. Is it *the* shortest? If it is the shortest road linking the four towns, how is it possible to prove that there is no shorter road system? Alternatively, if it is not the shortest system, what is the shortest roadway configuration and what is its length? This problem can be simply solved using an analogue method based on the minimization property of the area of a soap film.

3.3 The Analogue Solution

Before we attempt to solve the four town problem using an analogue method based on the use of soap films it will be instructive to attempt to obtain an analogue solution to the problem for which we know the analytic solution. That is, of course, the shortest path connecting two points; the straight line path. Consider a system consisting of two parallel clear perspex† plates separated by a distance b and connected by two pins perpendicular to the plates (Fig. 3.3). When this arrangement is immersed in a bath of soap solution and withdrawn from the bath a soap film will form between the plates. By symmetry it will be perpendicular to the plates, be bounded from above and below by the plates, and it will terminate at the two pins. Consequently it will have the same shape as a length of tape of width b. If l is the length of the tape of soap film, the area of the film is given by

$$A = bl. \tag{3.2}$$

This film is shown by the curved surface in Fig. 3.3. When the film comes to rest, in a state of equilibrium, A will be minimized. Hence, from Eq. (3.2), l will be minimized as b is a constant. That is, the length of the tape of soap film will be in the form of a straight line as shown in Fig. 3.3.

It is now clear how we should proceed in order to solve the four town problem. We must construct two parallel clear perspex plates joined by four pins, perpendicular to the plates, arranged at the corners of a square. This

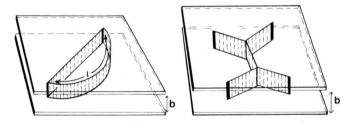

Fig. 3.3   Soap films, length l and width b, bounded by two parallel perspex plates and two pins. The curved film is a non-equilibrium film and the planar film is the equilibrium film.

Fig. 3.4   The equilibrium soap film configuration formed by four pins at the corners of a square contained between two parallel perspex plates.

represents, to scale, the four towns. The same argument as we developed for the two pin system will tell us that a soap film will form between the plates which has the form of a tape of constant width linked to each pin. It will

---

† In the United States of America this is known by the commercial name of *lucite*.

reach an equilibrium configuration in which the area, and hence the length, of the film will be a minimum. The final configuration is shown in Fig. 3.4 and Plate 3.0. The minimum length of road linking A, B, C and D contains two, three-way, intersections. Each intersection consists of three roads meeting at equal angles of 120°. Physically this is a consequence of the equilibrium of three equal surface tension forces. These forces act at each intersection in the directions of the films.

We can now calculate the length of this minimum roadway using the result that the angle between two soap films at an intersection is 120°. The total length is

$$1 + \sqrt{3} = 2.73.$$

This is approximately 4 per cent smaller than the value of 2.83 for the cross-roads system. There are two possible configurations with the same length. The second configuration can be obtained from the first by perturbing the film by blowing onto it. It is represented by the broken line in Fig. 3.5 and is equivalent to the initial configuration rotated through 90°. Table 3.1 contains all the roadway configurations examined and their respective lengths.

TABLE 3.1

The lengths of the motorway configurations.

Motorway Configuration	LENGTH	
	$4 + 2\sqrt{2}$	6·82
	$\pi\sqrt{2}$	4·44
	4	4·00
	3	3·00
	$2\sqrt{2}$	2·83
	$1 + \sqrt{3}$	2·73

It is now clear that we can extend this method to obtain the solution to the problem of linking n points by the smallest length of path. Consider five towns arranged at the vertices of a regular polygon. We must construct, to scale, a structure composed of five pins joining two perspex plates. Then immerse it in a bath of soap solution, withdraw it, and allow the system to drain. Once the soap film has come to rest the minimum roadway configuration is obtained (Fig. 3.6). Again, the area of the film will be minimized as the width of the film remains constant. The solution contains three intersections. Each intersection consists of three intersecting roads and the angle between adjacent roads is 120°.

It is a general property of the solutions to these problems that they consist of straight lines forming a number of intersections. These intersections always contain three straight lines with adjacent lines intersecting at 120°. The number of intersections, in the case of n points, will be in the range zero to a maximum of $(n-2)$.

At the points where the pins are situated it is possible for one, two, or three lines to meet. If two lines meet at such a point the angle between the lines must be greater or equal to 120°. If the angle is less than 120° it would be possible for the two lines to coalesce and form a configuration with only one line linked to the point. This line would be one of the roads radiating from a three way intersection (Fig. 3.7) and would give rise to a shorter total length

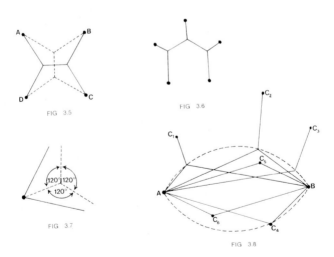

FIG 3.5

FIG 3.6

FIG 3.7

FIG 3.8

Fig. 3.5 The two minimum paths of equal length linking A, B, C and D.
Fig. 3.6 The minimum configuration linking five towns at the vertices of a regular pentagon.
Fig. 3.7 The configuration formed by the broken lines has a smaller total length than that formed by the full lines.
Fig. 3.8 Minimum paths linking fixed points A and B, and a variable point C.

than the system of two lines. In the special case in which a point occurs at a three-way intersection it is possible for three lines to meet at a point.

It is worth examining these results in greater detail. Consider the three point, or three town, problem. Let A and B be fixed points and C a variable point. Frequently the minimum path linking A, B and C will consist of three lines meeting at a three-way intersection with $120°$ angles. Examples are given in Fig. 3.8 of different positions of C, C_1, C_2 and C_3, which give rise to a three-way intersection. In each case there is only one line connected to C. As the position of C changes the intersection will move along the broken curves. The intersection point in these cases subtends an angle of $120°$ at A and B. This angle remains constant as C is varied, consequently the dotted curves must be arcs of circles. This follows from the converse of the geometrical theorem that states that angles in the same segment of a circle are all equal. The upper and lower broken arcs belong to different circles.

If C lies on one of the broken arcs there will only be two straight lines forming the minimum path, AC and CB. In Fig. 3.8 this occurs when C is at position C_4. The angle $A\hat{C}_4B$ will be $120°$, and the single line joining the intersection to C will have been reduced to zero length. When the point C lies inside the dotted arcs the angle $A\hat{C}B$ will be greater than $120°$ and there will be two lines joined to C. C_5 and C_6 are examples of this situation.

3.4   Relative and Absolute Minima

There is only one configuration with the property of minimum length for the two point problem and the three point problem. Consequently if the shape of the path connecting the points is altered by means of a parameter p, or parameters p_1, p_2 . . . etc., so that it no longer has its minimum length, the configuration can be made to change continuously. There will be one value of p, or p_1, p_2 . . . etc. for which the configuration with the minimum length will be obtained. In the case of the two point problem an example of such a parameter might be the radius of a circle passing through the two points, the path linking the points being the arc of the circle. As the radius is increased the length of arc joining the points will decrease and pass through its minimum value when the radius becomes infinite. That is, when the curve joining the two points is a straight line. As the radius of curvature changes sign and decreases in magnitude, the length of arc will increase in magnitude. Figure 3.9 shows the variation of the length of path $L(p)$ as a function of p. The minimum value is labelled p_M. The problems with four points at the corners of a square and five points at the corners of a regular pentagon also have only one basic minimum configuration. There is more than one minimum configuration in both these cases. However the different configurations can be obtained by rotating a minimum configuration about the centre of the polygon. That is, they are degenerate as they are identical when rotated; all having the same length. For example in the case of the square a rotation through 90° (Fig. 3.5) will give the other minimum. Similarly for the pentagon, a rotation through multiples of 72° will give all the minimum configurations, each having the same length. For all these problems there is one basic minimum configuration.

It is possible to conceive of situations for the general problem of n points where the function $L(p)$, or $L(p_1, p_2 \ldots)$, will have more than one minimum. Each minimum length being associated with a different configuration. In this case we will require the smallest minimum, or *absolute* minimum. Physically the absolute minimum gives the soap film configuration of lowest energy, and hence smallest area. The other minima will be *relative* minima, as they correspond to metastable equilibrium configurations. Figure 3.10 shows a typical $L(p)$ versus p curve for the general problem. M_1, M_2, $M_3 \ldots$ etc., label the minimum values of p. It is seen that M_2 gives the absolute minimum for the situation in Fig. 3.10.

An example of a problem with several minimum configurations occurs in the determination of the minimum path linking six points forming the ver-

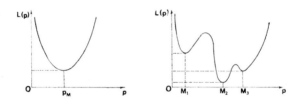

Fig. 3.9   The path length, $L(p)$, as a function of parameter p. p_M is the value of p for which $L(p)$ is a minimum.
Fig. 3.10   The variation of $L(p)$ with p for the case in which $L(p)$ has several minima.

tices of a regular hexagon. In this case there are three minimum configurations. They can all be obtained by using the procedure of constructing parallel plates joined by six pins arranged at the corners of a regular hexagon, to represent a scale model of the problem. It is now possible to form three different equilibrium configurations of the soap film. Each configuration corresponds to a minimum value of the length of the path joining the six points. They can each be obtained by withdrawing the plates from the soap solution, using different angles of withdrawal to obtain each configuration. The three configurations are shown in Fig. 3.11 and in Plates 3.1. Alternatively, having obtained one of the minimum configurations it is possible to perturb the film by blowing onto it so that it jumps into another equilibrium configuration. All three configurations can be obtained in this way.

Figure 3.11(b) has the highest rotational symmetry. It has a three-fold axis of rotation about the centre of the hexagon. Figure 3.11(c) has a two-fold axis of symmetry, and Fig. 3.11(a) a one fold-axis of symmetry about the centre of the hexagon. To obtain the configuration with the absolute minimum of length, it is necessary either to calculate the lengths of all the minimum paths in Fig. 3.11, using the information that the intersections contain 120° angles, or to measure the lengths of the three paths. The smallest path gives the absolute minimum. The absolute minimum in this case is given by Fig. 3.11(a).

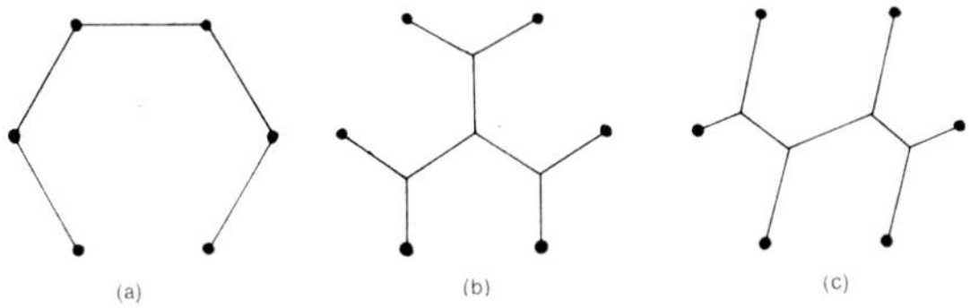

Fig. 3.11   The three minimum configuration for the regular hexagon of points; (a) has a one-fold rotational axis of symmetry; (b) has a three-fold rotational axis of symmetry; (c) has a two-fold rotational axis of symmetry, (a) has the smallest path length.

For a regular polygon with six or more sides the absolute minimum is the circumferential path with one of the sides absent. By omitting one of the sides of the polygon it is still possible to link all vertices of the polygon.

Fig. 3.12   A minimum roadway linking a number of towns in the British Isles.

This analogue technique can easily be applied to problems containing any number of points, or towns. It has applications to the determination of the minimum length of road, electricity cable, gas pipes, etc., linking towns. At present the only result that is widely used is that concerning the shortest path connecting two points; the straight line. In roadway construction the road usually deviates from the straight line to take account of geographical constraints that would increase the cost of construction appreciably, such as mountains and lakes. Also sociological constraints, such as population centres, may also require the road to avoid a particular region. However the starting point for the suggested path of the road is based on the solution to the problem of the shortest distance between two points. It would be useful for planners to be familiar with the more general results concerning the shortest path linking a number of points when, initially, attempting to link several towns by roads or pipes.

An example of a minimum path linking a number of towns in the British Isles is shown in Fig. 3.12. In order to obtain this minimum path a map of the British Isles has to be drawn to scale on one of the perspex plates. Pins must be inserted at the positions of the towns that are required to be linked. In order to find the shortest path one has to exhaust all the possible minimum paths and hence determine, by measurement or calculation, the smallest minimum path.

3.5   The Curvature of the Earth

All of the previous problems concern the determination of the minimum length of path joining a number of points in a plane. In the case of problems requiring the minimum distance joining a number of towns we have assumed that the distances are small compared with the radius of the earth. Under these conditions the towns can all be considered to be in a plane. When this approximation is not justified it is necessary to take account of the curvature of the earth. This can easily be done by extending the techniques that we have developed. However, instead of using parallel flat plates we require concentric spherical shells. The towns are represented again by pins perpendicular to the plates, joining the two spherical shells. They represent, to scale, the towns on the surface of the earth. The surface of the earth can be taken as either spherical shell. Part of a spherical shell must be cut away so that soap solution can enter between the plates. When the system is dipped into a bath of soap solution and withdrawn, a film will form between the two spherical shells and link all the pins. By symmetry the film will be perpendicular to the surfaces of the shells. Each portion of film contained by the concentric spherical shells will be in the shape of planar sections formed between two concentric coplanar circles (Fig. 3.13). If the radii of the shells are R_1 and R_2

and $R_1 > R_2$, then the area of a section of film of length l_i, as measured along the surface of the larger shell, is

$$\tfrac{1}{2}(R_1{}^2 - R_2{}^2)\frac{l_i}{R_1}. \tag{3.3}$$

Adding up the area of all such sections the total area, A, for N sections of film ($i = 1, 2, \ldots N$) is given by

$$A = \left(\frac{R_1{}^2 - R_2{}^2}{2R_1}\right) \sum_{i=1}^{N} l_i, \tag{3.4}$$

$$= \left(\frac{R_1{}^2 - R_2{}^2}{2R_1}\right) L, \tag{3.5}$$

where L is the total length of soap film measured on the larger spherical shell. Hence A is proportional to the total length of film, L. So, as in the case of the parallel plates, the area of the soap film is proportional to the length of the film joining the pins. When the film comes to rest it will have its minimum area and hence give the minimum path joining the pins. This path is measured along the surface of the larger shell. In the case of only two pins the length of path corresponds to the geodesic curve linking the extremities of the pins by the shortest path.

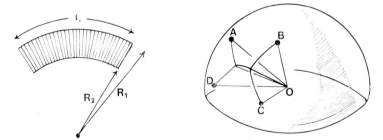

Fig. 3.13   A section of soap film, of length l_i, contained by two spherical shells of radii R_1 and R_2.

Fig. 3.14   The minimum path formed by four towns A, B, C and D on the surface of the earth. A single spherical shell and radial wires are used to form a soap film with minimum area and hence minimum path.

In practice it is convenient to use only one spherical shell. This is possible by making $R_2 = 0$. Thus one constructs straight wires from the centre of the pherical shell to the points on its surface, which represent the towns on the surface of the earth. The towns should be situated, to scale, at the correct latitudes and longitudes. After withdrawing the system from a bath of soap solution the soap film, bounded by the wires and the shell, will consist of radial sectors. The total length of these, measured on the surface of the shell, will give the minimum path connecting the towns once the soap film has come to rest. The solution to the minimum path connecting four towns on the surface of the earth is shown in Fig. 3.14 and Plate 3.2(a). Once again there are two three-way intersections and the angles of adjacent paths, measured on the surface of the spherical shell, is 120°. The angle between the planes of soap film meet-ing along lines of soap solution radiating from the centre of the shell are also 120°. This technique can be applied to any number of points on the surface of the shell. In Plate 3.2(b) the solution for three points is given. It will possibly find applications in the building of the next generation of motorways linking continents.

The technique of constructing a structure composed of wires joined to a surface can also be applied to the solution of problems requiring the mini-mization of the path connecting a number of points in a plane. For example the four town problem in a plane can be solved by constructing four wires radiating from any point at any height, h, above a perspex plate to the four points on the plate that represent, to scale, the four towns. After the system is withdrawn from a bath of soap solution a surface of soap film will be formed bounded by the wires and the perspex plate. The film will be composed of sections of film that are perpendicular to the plate. Consequently they are a series of triangular planar soap films all of height h and with their bases in the plane of the perspex plate (Fig. 3.15). The total area of the soap film, A, in the case of N triangular sections is thus given by,

$$A = \tfrac{1}{2}h \sum_{i=1}^{N} l_i, \tag{3.6}$$

where l_i is the length of the base of a triangular soap film in the plane of the plate, $i = 1, 2, \ldots N$.

If L is the total length of all the bases of the triangular soap films,

$$A = \tfrac{1}{2}hL. \tag{3.7}$$

Thus A is proportional to L. The soap film will take up a configuration in which the area is minimized once the film has come to rest. Thus, from (3.7), L will be minimized. So the minimum length of path connecting the four points is obtained. This result is obviously valid for any number of points.

3.6   Constraints

So far no attempt has been taken to solve problems in which the mini-mum path linking a number of points must be constrained to avoid certain regions. These constraints can arise in roadway problems from hills and lakes. It is difficult, and expensive, to build roads over hills and across lakes. Con-sequently the roadway must often avoid regions where a hill or lake will make road construction difficult. Other constraints prevent roads being built through historic areas, areas of natural beauty and through existing resi-dential areas. How can these constraints be taken into account when solving problems?

A simple example will illustrate how we can solve problems with con-straints using soap films. Consider the problem of three towns with a circular lake at the centre (Fig. 3.16). It is required to build a road linking the towns by the shortest length of road which avoids the construction of a bridge over the lake. To solve this problem we construct two parallel perspex plates con-nected by three pins to represent, to scale, the towns. A circular hole is drilled through the plates to represent, to scale, the lake. If the plates are now dipped into soap solution, a soap film will form between the plates when they are withdrawn from the solution. The soap film can be prevented from forming inside the lake, between the plates, by breaking any film that forms there. The soap film will come to rest and form a configuration that joins the three pins and avoids the hole representing the lake. The film will be perpendicular to the plates in the form of a tape. Its area will be proportional to its length. The equilibrium configuration of the film will be that of minimum area and hence of minimum length. Thus the minimum length of path joining the three towns that avoids the lake can be obtained. It consists of a three-way inter-

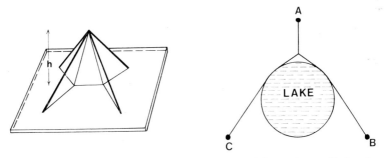

Fig. 3.15   A system consisting of a horizontal plate and four wires radiating from a point above the plate can be used to solve the problem of the minimum path joining the four points at the base of the wires. The length of the soap film measured in the horizontal plane of the plate gives the minimum length of path.

Fig. 3.16   The minimum path linking *A*, *B*, and *C* with a circular constraint.

section (Fig. 3.16 and Plate 3.3) and avoids the lake by skirting around part of the circumference.

This method can be applied to constraints with boundaries of any shape. It is only necessary to drill holes in the plates to represent, to scale, the constraints. It is not necessary to drill holes in both plates, one plate is sufficient. In the map of the British Isles (Fig. 3.12) account could be taken of the Pennines by drilling a hole in the plates and preventing any film from forming inside the hole.

When the curvature of the earth has to be taken into account it is necessary to drill a hole, to scale, in the spherical shell. The soap film must be prevented from entering this region. In the most general case all the minimum configurations have to be obtained and the absolute minimum determined.

These minimization problems were first investigated early in the nineteenth century by the mathematician Jacob Steiner[14, 26]. They are often known collectively as the Steiner problem. The general problem of linking n points has not been solved analytically. It was the mathematician Richard Courant[14, 15] who first popularized the analogue solutions to these problems in the 1940's.

3.7 Practical Considerations

In describing the analogue systems of plates and pins, or wires, it has been assumed that the diameters of pins are infinitely small. In practice the diameters will be small, but finite. The finiteness of the diameter can produce results that are different from those for an infinitely small diameter. However by being aware of these differences it is possible, when necessary, to distort the shape of the soap film around the pins so that it has the same configuration as that for a wire of infinitely small diameter.

The finiteness of the pin diameter can result in two soap films being attached to a pin at an angle of less than 120° (Fig. 3.17(a)). We know that this is impossible for a pin of infinitely small diameter. The films should coalesce to produce a three-way intersection with only one film attached to the pin. This configuration can be produced by perturbing the films joined to

(a)   (b)

Fig. 3.17   (a) A possible soap film configuration formed at a pin of finite diameter. (b) The configuration required to solve the minimum path problem for the case of pins with zero diameter.

the pin by blowing onto them so that they coalesce. The soap film configuration will now be identical to that for an infinitely small diameter pin (Fig. 3.17(b)). All measurements of the length of film attached to the pins should extend to the central axis of the pins.

No special soap solution is required for these demonstrations. A bowl of cold water plus a couple of squirts of any washing-up liquid is adequate providing the solution is stirred and bubbles are removed from the surface of the liquid. The films between a set of plates will last for about 60 seconds. For longer lasting soap films one should prepare one of the special solutions mentioned in section 1.7.

If one holds a finger or hand across the two parallel plates it will act as a pin and a soap film will become attached to it after withdrawal from the soap solution. So one should always take care to hold the system by only one of the plates when withdrawing it from the bath of soap solution.

It is sometimes convenient to project the soap film configuration onto a screen. This is easily done by placing the plates on an overhead projector. The films show up as black lines. It is also possible to change the soap film configuration between the plates by blowing onto it whilst it is on the overhead projector. It will be seen to jump from one minimum configuration to another. In this way the configurations can be viewed by a large audience.

3.8 A Geometrical Proof of the Three Point Steiner Problem

All the problems discussed in this chapter have been solved by analogue methods based on the minimization property of the surface area of soap films. We shall now solve using analytic means one of the simpler problems, the problem of finding the minimum path linking three points.

In order to prove the general result for the minimum path linking three points it will be convenient to recall some properties of the ellipse. If an ellipse has foci A and B (Fig. 3.18), then any point, P, on the ellipse has the property that

$$AP + PB = l, \tag{3.8}$$

where l is constant. Also AP and PB have the property that they are equally inclined to the tangent to the ellipse at P, TT'. They are also equally inclined to the normal at P, NPN', so that

$$T\hat{P}A = T'\hat{P}B,$$

and

$$A\hat{P}N = B\hat{P}N,$$

or

$$A\hat{P}N' = B\hat{P}N'. \tag{3.9}$$

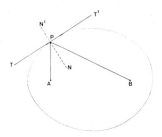

Fig. 3.18   An ellipse with foci A and B.

Confocal ellipses with foci at A and B, having axes of increasing length, (Fig. 3.19) will have increasing values of l. In triangle P_1P_2B (Fig. 3.19) the sum of two sides of the triangle are greater than the third, so

$$P_1P_2 + P_2B > P_1B. \qquad (3.10)$$

Adding AP_1 to both sides of (3.10),

$$AP_1 + P_1P_2 + P_2B > AP_1 + P_1B. \qquad (3.11)$$

That is

$$AP_2 + P_2B > AP_1 + P_1B. \qquad (3.12)$$

Now let us consider the problem of finding the minimum path joining three points, A, B and C. Let P be the required point of intersection of the straight line paths from A, B and C (Fig. 3.20) that constitute the minimum

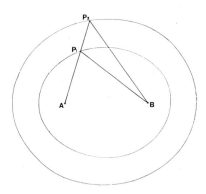

Fig. 3.19   Confocal ellipses.

path linking *A*, *B* and *C*. The minimum path obviously cannot consist of curved lines as any curved line can be replaced by a straight line of shorter length. It is possible that *P* may coincide with one of the vertices of triangle *ABC*. If this is the case it must coincide with the vertex with the largest angle, say *C*, as the total path (*AC*+*CB*) is less than the sum of any other pair of sides of the triangle. We must now investigate the case in which *P* *does not* coincide with a vertex, and also discover the conditions under which *P* coincides with a vertex.

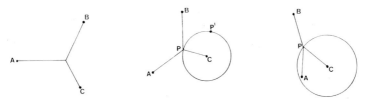

Fig. 3.20   The three points *A*, *B* and *C* are joined to the point *P* to give the minimum total path.

Fig. 3.21   *P'* lies on a circle radius *CP*.

Fig. 3.22   *A* lies inside the circle radius *CP*.

Consider a circle centre *C* and radius *CP* (Fig. 3.21). *Q* is any point and the point *P* is assumed to minimize (*AQ*+*BQ*+*CQ*). Thus any point on the circumference of the circle radius *CP* such as *P'* (Fig. 3.21) must satisfy

$$AP'+BP'+CP' \geqslant AP+BP+CP. \tag{3.13}$$

As *CP'* = *CP*, both lengths being radii of the circle,

$$AP'+BP' \geqslant AP+BP. \tag{3.14}$$

Thus (*AP'*+*BP'*) must have the smallest possible value when the minimum path is obtained.

Consider a set of confocal ellipses with foci at *A* and *B* that intersect the circle. The ellipse that touches the circle will have the smallest value of the constant '(*AP'*+*BP'*)'. Any ellipse that intersects the circle in two different points will have a greater value of this constant, Eq. (3.12). Thus *P* must coincide with the point at which the ellipse touches the circle. For this ellipse, from (3.9),

$$A\hat{P}C = B\hat{P}C, \tag{3.15}$$

as *PC* is normal to the tangent at *P*.

The same argument can be applied to the circle centre A and radius AP to give,

$$A\hat{P}B = A\hat{P}C. \qquad (3.16)$$

From (3.15) and (3.16)

$$A\hat{P}C = A\hat{P}B = B\hat{P}C. \qquad (3.17)$$

These angles must add up to $360°$ and are thus all equal to $120°$. P is the point that subtends an angle of $120°$ with any pair of points A, B and C.

These results have assumed that A and B lie outside the circle centre C for the case in which P does not coincide with a vertex. If at least one of the points, say A, were inside or on the circle centre C, then from Fig. 3.22,

$$AP + BP \geqslant AB, \qquad (3.18)$$

as the sum of two sides of triangle APB must be greater than, or equal to, the third. Also

$$CP \geqslant AC, \qquad (3.19)$$

as CP is the radius of the circle and the point A lies inside the circle. Adding (3.18) and (3.19),

$$AP + BP + CP \geqslant AB + AC. \qquad (3.20)$$

This implies that the shortest path $(AP + BP + CP)$ will be obtained if P coincides with A. This is contrary to our assumption that P does not coincide with a vertex. This conclusion also applies when both A and B are assumed to lie inside the circle. So A or B, or both, cannot lie inside the circle. Consequently they both lie outside the circle. These conclusions also apply to circles with centres A and B.

3.9   Analysis of Solutions

We have shown that the minimum path linking A, B and C can have one of two possible configurations: either it consists of the two sides of triangle ABC adjacent to the largest angle of the triangle, or it consists of three straight lines from each of the points A, B and C to a point P such that $A\hat{P}B = B\hat{P}C = C\hat{P}A = 120°$. It is necessary to determine the conditions for each solution.

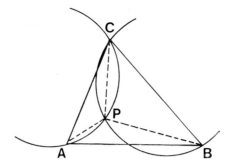

Fig. 3.23   *P* lies on the minor arcs *AC* and *CB* if the triangle *ABC* has no angle greater than 120°

It will be useful to use the geometrical theorem that angles in the same segment of a circle are equal. The point *P* lies on the minor arc of the circle through *A* and *C* which subtends angles of 120° at *A* and *C*. It also lies on the minor arc of the circle through *C* and *B* which subtends angles of 120° from *A* and *B*. *P* thus lies at the intersection of the two circles (Fig. 3.23) at which

$$A\hat{P}C = B\hat{P}C = 120°. \tag{3.21}$$

If triangle *ABC* has no angle greater than 120° the circles intersect inside the triangle and *P* will lie inside the triangle. If one of the vertices, say *C*, has an angle greater than 120° (Fig. 3.24) the circles will intersect outside the tri-

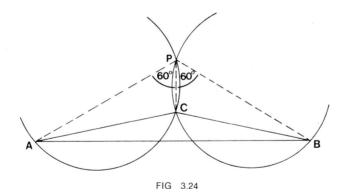

FIG 3.24

Fig. 3.24   When *ACB* >120° there is no point *P* inside triangle *ABC*.

angle on the major arcs of the circles. The point of intersection, *P*, no longer subtends an angle of 120° as it now lies on the major arc. It subtends angles of 60°. In this case there is no point that subtends an angle of 120° with all pairs of vertices of the triangle. Consequently the only alternative solution is that the minimum point, P, must coincide the vertex of the obtuse angle of the triangle.

To summarize; a minimum path can be constructed such that P subtends angles of $120°$ with each side of the triangle providing the angles of triangle ABC are less than $120°$. If one of the angles of the triangle is greater than $120°$ the minimum path consists of two sides of the triangle adjacent to the obtuse angle. There is also the possibility in the former case that the path is greater than that formed by the two adjacent sides of the triangle, where P coincides with the vertex having the greatest angle. Thus it remains to eliminate this possibility.

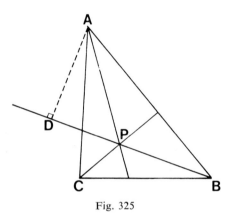

Fig. 325

In order to complete the proof we shall extend BP to meet the perpendicular from A at D (Fig. 3.25) for the case of a triangle in which no angle is greater than $120°$. Then as $B\hat{P}A = 120°$,

$$D\hat{P}A = 60°. \tag{3.22}$$

In triangle PAD, $\hat{D} = 90°$ and $\hat{P} = 60°$, thus

$$PD = \tfrac{1}{2}PA, \tag{3.23}$$

$$\therefore BD = \tfrac{1}{2}PA + PB. \tag{3.24}$$

Now in triangle BDA, $\hat{D} = 90°$, so

$$BD < AB. \tag{3.25}$$

Substituting from (3.24) into (3.25),

$$\tfrac{1}{2}PA + PB < AB. \tag{3.26}$$

Similarly by dropping a perpendicular from A onto the extension of CP,

$$\tfrac{1}{2}PA + PC < AC. \tag{3.27}$$

Adding (3.26) and (3.27),

$$PA + PB + PC < AB + AC. \tag{3.28}$$

Similar inequalities exist in which the right hand side of (3.28) consists of any two sides of the triangle ABC. We have thus shown that it is not possible for any two sides of the triangle to be smaller than $(PA + PB + PC)$ for a triangle containing no angle greater than $120°$. This proves the result first obtained by analogue methods, in section 3.3, using soap films.

It is possible to extend these results to problems with more points to obtain minimum path configurations with the properties obtained above. However, no explicit general proof is available of the path configuration and length for linking n points by the minimum path. The intersections with three lines meeting at $120°$ are characteristic of the general problem. The maximum number of intersections required to link n points is $(n{-}2)$. However the number of intersections may take a value in the range 0 to $(n{-}2)$.

Gilbert and Pollak[90] have shown that it is possible to write down a simple expression for the number of minimum paths, of different topological types, linking the n points. The length of these paths can be evaluated using a digital computer, and the absolute minimum path determined providing the value of n is within the computing capability of the machine.

4 MINIMUM SURFACES
IN THREE DIMENSIONS

4.1 Three Dimensional Problems

In the previous chapter we applied the minimum area property of soap films to the solution of some two dimensional problems requiring the minimization of the path length linking a number of points. This was carried out by using two parallel plates in order to maintain a constant film width. In this chapter this constraint will no longer be imposed and more general three dimensional minimization problems will be examined.

A soap film contained by any fixed boundary will acquire its minimum free energy when it reaches equilibrium. As the free energy of a film is proportional to its area, the area will also be minimized. Consequently soap films can be used to solve mathematical problems requiring the minimization of a surface area contained by a boundary. In order to obtain analogue solutions we require a frame to form the boundary of the surface. When the frame is withdrawn from a bath of soap solution a soap film will form which will attain its minimum area configuration on reaching to equilibrium.

4.2 Classification of Surfaces in Three Dimensions

There are certain properties of surfaces that remain unchanged if the surface is distorted. These are called topological properties. We can classify surfaces with different characteristic topological features. Examples of two surfaces with different topological features are the sphere and the donut, or torus (Fig. 4.1). If we draw a closed curve c_1 on the surface of the sphere it will isolate the surface inside c_1 from that outside and hence we have two separate regions. However if we draw a closed curve c_2 around the torus (Fig. 4.1(b)) we still have only one surface. Other surfaces can be constructed, such as that shown in Fig. 4.1(c), which have additional 'handles' and contain more closed contours while still retaining a single surface. Figure 4.1(c) has a

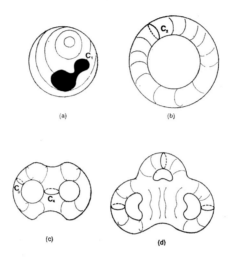

Fig. 4.1   Different topological surfaces: (a) has genus zero; (b) has genus one; (c) has genus two; (d) has genus three.

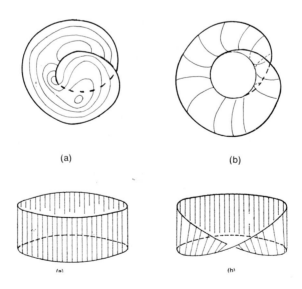

Fig. 4.2   Different minimum surfaces: (a) has twos urfaces; (b) has only one surface.
Fig. 4.3   (a) A film with two surfaces. (b) A Moebius strip of soap film.

maximum of two such closed contours, c_3 and c_4. It is convenient to distinguish the different topological properties by defining a surface property called the *genus*. This is the largest number of non-intersecting simple closed curves that can be drawn on the surface without separating it into two distinct regions. The genus of the sphere is thus zero, that of the torus one, that of the surface in Fig. 4.1(c) two, and that in Fig. 4.1(d) three, etc. A surface with p holes, or handles, will have a genus of p. The definition has been obtained by considering two sided surfaces.

4.3 Topologically Different Soap Film Surfaces

The American mathematician Richard Courant[14, 15] has drawn attention to the topologically different surfaces that can be obtained with wire frames and soap solution. Altering the shape of the wire boundary can change the topological nature of the equilibrium surface of the soap film. Also, by causing the soap film to change from one minimum surface to another one may change the topological character of the surface. This change can be induced by blowing onto the film or by shaking it.

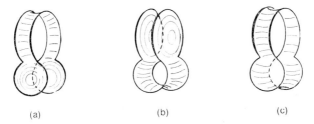

(a)   (b)   (c)

Fig. 4.4   Different minimum surfaces produced by a wire frame in the shape of earphones: (a) and (b) have genus zero; (c) has genus one.

Consider a soap film contained by a wire ring. It will have two surfaces and have genus zero. If the ring is distorted the soap film will take up the surface with minimum area determined by the new boundary, such as that shown in Fig. 4.2(a). This retains both surfaces but can jump into another minimum area configuration with only one surface (Fig. 4.2(b)). It has the same topology as a Moebius strip. A simple Moebius strip can be produced by using two coaxial rings of wire (Fig. 4.3(a)). If a cut is made in both rings and they are joined so that the upper one is connected to the lower one, as shown in Fig. 4.3(b), the soap film bounded by the wire will take up the shape of a Moebius strip as indicated in Fig. 4.3(b).

Another interesting example is obtained with a wire boundary in the shape of a pair of earphones (Fig. 4.4). There are three possible equilibrium configurations for the soap film (Figs. 4.4(a), (b), and (c)). Figures (a) and (b) have a genus of zero and (c) has a genus of one. The absolute minimum area is not immediately obvious but can be determined, with difficulty, by measurement. The surface can jump from one shape to another by blowing onto the soap film, or by shaking it.

4.4 Some Problems with an Analytic Solution

Most of the problems requiring the minimization of the area contained by a fixed boundary have not been solved analytically by mathematicians. However a few have been susceptible to analytical methods and these will be examined here.

Leonard Euler[41] in 1744 obtained a differential equation for determining the minimum area contained by a fixed boundary. His analysis was later simplified by Lagrange.[101] These results arose out of earlier studies of minimization problems by distinguished mathematicians[55] such as Newton and the Bernoullis. In fact Euler was a student of John Bernoulli, who together with several contemporary mathematicians solved one of the most challenging minimization problems of his day, the brachistochrone problem. This problem concerns the determination of the path taken by a particle falling in a gravitational field from point A to point B such that the time taken is minimized. John Bernoulli showed that the path is a cycloid, which is the curve produced by a point on the circumference of a wheel that rolls, without slipping, along a straight line (Fig. 4.5). The general equation derived by Euler and Lagrange, that can be applied to the minimum area problems, is obtained in Appendix I. It assumes that the solutions to minimum area problems are continuous and have continuous derivatives. That is, the surface is smooth, does not contain any discontinuities, and is without discontinuities in the gradient of the surface.

Fig. 4.5   A cycloid produced by the motion of a point on the circumference of a wheel that rolls without slipping.

In the Steiner problems, examined in Chapter 3, the assumption concerning continuous gradients is not justified. Most of the solutions contain intersections with gradients that change discontinuously. This will also be true of the majority of the problems to be discussed in section 4.5 and so it will not be possible to apply the Euler–Lagrange equations.

The methods of Euler and Lagrange consider a continuous sequence of surfaces, contained by the boundary, which deviate from the extremum surface. For the soap film problem the extremum surface is the minimum area surface. A differential equation is then derived for the extremum surface. The solution to the differential equation will give the required surface.

In Appendix II the Euler–Lagrange equation is applied to the well-known problem of the determination of the minimum path between two points. The solution gives the straight line path between the two points. As soon as one tries to tackle more difficult problems the solution of the differential equation rapidly increases in difficulty. It has been thought worthwhile to include a further example to illustrate some of these difficulties. This is the problem of the minimum area contained by two coaxial parallel rings of equal diameter. The complete solution to this problem is given in Appendix III. It is shown that the minimum surface is a catenoid, or catenary of revolution, bounded by the two equal rings, providing that the rings are sufficiently close together. A catenary is the curve taken up by a hanging chain. As the distance between the rings is increased the catenoid dips further towards the axis (Fig. 4.6). The area of the catenoid will be the absolute minimum if the distance between the rings is less than 1.056a, where a is the radius of the rings. It will be a relative minimum for distances between 1.056a and 1.325a. For distances greater than 1.325a Goldschmidt[92] showed analytically in 1831 that the surface jumps discontinuously into the two discs, each disc being contained by a ring. For distances greater than 1.056a the two discs give the surface with the absolute minimum area. The solution with surfaces that are discontinuous has become known as the Goldschmidt discontinuous solution.

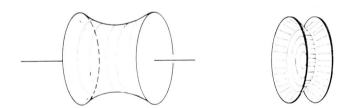

Fig. 4.6   A catenoid produced by a soap film contained by two rings.

Fig. 4.7   Another minimum surface contained by two rings, containing an intermediate disc.

These minimum surfaces can easily be obtained using soap films. When two rings are dipped into soap solution a continuous soap film, contained by the two rings, can be produced. It will have its minimum area once it has come to rest. If the rings are initially close together one obtains a surface in the form of a catenary of revolution. As the distance between the rings is increased the film will dip further towards the axis. At a separation of 1.325a the film will jump, discontinuously, from the catenoid into the two discs. Further separation of the rings will maintain this absolute minimum surface.

There is another surface that provides a solution to the minimum area contained by two rings. This is an axially symmetric surface, contained by the two rings, with a vertical disc placed symmetrically between the two rings. It is shown in Fig. 4.7. At the circumference of the central vertical disc the gradient changes discontinuously. The circumference of the disc is formed by the intersection of three surfaces. The three tangent planes to these surfaces, at any point along the circumference, intersect at $120°$ to each other. The Euler–Lagrange equation is not applicable to this problem as the surface contains discontinuities in the gradients along the circumference of the central disc. An analogue solution to the minimum area problem can be produced by dipping two rings into a bath of soap solution and obtaining the equilibrium surface. By breaking the central disc of soap film the catenary of revolution, or catenoid, (Fig. 4.6) is obtained. When producing the catenary of revolution it is necessary to ensure that the central disc is not formed, or to break it if it has formed.

The solution to the problem of the minimum area contained by two rings can be extended to the case of two rings that are no longer coaxial yet remain parallel (Fig. 4.8). As the rings are moved sideways relative to each other, keeping the distance, h, between the planes of the rings fixed, the surface becomes a distorted catenoid. The distortion increases with the relative distance between the rings until the horizontal distance between the centres

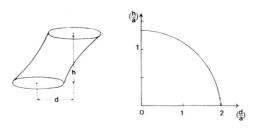

Fig. 4.8   The minimum surface contained by two parallel horizontal rings with centres separated by a horizontal distance, d, and by a vertical height h.

Fig. 4.9   The variation of h with d for the limiting surface joining the two rings. Outside this curve the Goldschmidt discontinuous solution is obtained.

of the rings, d, lies on the curve shown in Fig. 4.9 of h against d. Beyond this value of d one obtains the Goldschmidt discontinuous solution. If the horizontal distance, d, is greater than the diameter of the ring the only possible solution for any separation, h, is the Goldschmidt discontinuous solution consisting of two discs bounded by the rings. The graph (Fig. 4.9) will have $h = 1.325a$ when $d = 0$, which is the value for the coaxial problem.

Another variation on this problem can be produced by using two conical funnels to bound the soap film surface. Figure 4.10 shows an axially symmetric surface joined to the rim of the smaller funnel and free to move inside the surface of the larger, coaxial, funnel. At equilibrium the surface will, by symmetry, make an angle of contact of 90° with the internal surface of the larger funnel as the funnel can only exert a force normal to its surface to balance that due to film tension. This surface can be created by dipping the two funnels into the bath of soap solution. The free boundary of the soap film, making contact with the surface of the larger funnel, will move to a position that minimizes the area of the film for any separation of the funnels. The soap film will produce a surface with the shape of a catenoid once it has reached equilibrium. As the funnels are moved apart the film will slide up the larger funnel until the separation is reached at which the film becomes unstable and breaks up into the Goldschmidt discontinuous surfaces.

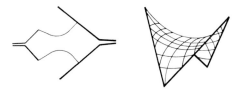

Fig. 4.10   The minimum, catenoid, surface contained by two coaxial funnels.

Fig. 4.11   The minimum surface contained by a skew quadrilateral.

At the end of the last century H. A. Schwarz[23] solved analytically another problem in which the minimum surface is continuous and has continuous derivatives. This problem concerns the minimum surface area contained by a skew quadrilateral. That is a quadrilateral which is not confined to a plane. The solution can be expressed in terms of hyperelliptic integrals and has the shape of a saddle (Fig. 4.11). The analogue surface can be obtained by forming a soap film with a boundary formed by a wire frame which has the shape of the skew quadrilateral.

4.5   Analogue Solutions to Some Unsolved Problems

Let us now examine some minimum surfaces that have not been solved analytically. In Chapter 3, when discussing the Steiner problem, we showed that it was possible to deduce some general geometrical properties of soap films by examining the minimum path configurations. We would expect to be able to deduce generalizations of these results in three dimensions. These results could provide a guide to the analytic solutions to minimum area problems. The best choice of frameworks for bounding the soap film are perhaps those with the highest symmetry. So the initial investigations should be concerned with the frameworks in the shape of the Platonic solids; the tetrahedron; the cube; the octahedron; the dodecahedron; and the icosahedron. These are shown in Fig. 4.12(a, b, c, d, e). The Platonic figures have regular faces, all congruent, with equal face-angles at every vertex and all the angles between adjacent faces, the dihedral angles, are equal.

One might conjecture that the minimum surface area bounded by the six edges of the tetrahedron is made up of three triangular faces of the tetrahedron. The fourth face would not be required as the surfaces formed by the three faces are joined to all six edges. Further thought suggests that a smaller area could be obtained by pinching together part of two adjacent surfaces (Fig. 4.13). There would then be an area common to the two faces, two

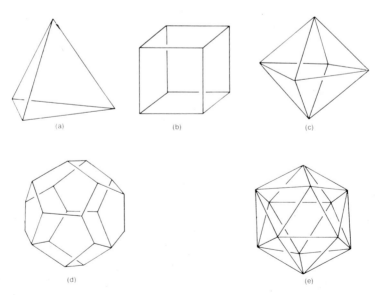

Fig. 4.12   The Platonic frameworks: (a) tetrahedron; (b) cube; (c) octahedron; (d) dodecahedron; (e) icosahedron.

sections of the original two faces would be merged into one surface, thus reducing the area. One could extend this procedure to all sets of neighbouring faces until the minimum area was obtained. It is not, however, clear what the shape of the final surface will be. One way to obtain the minimum surface is to dip a tetrahedral framework into a bath of soap solution. The minimum surface is formed after the framework has been withdrawn and the soap film has come to equilibrium. Plate 4.1 shows this final surface. Interference effects produce the horizontal colour fringes to be seen in the surface. The surface consists of six planar surfaces, each being in the shape of an isosceles triangle and bounded by an edge of the tetrahedron. All the surfaces meet at the centre of the tetrahedron. Each of the two internal edges of a triangular section of soap film is joined to two other surfaces. The four lines of soap film form the internal edges of the triangular surfaces, and each line connects a vertex of the tetrahedron to its centre. These lines are each formed by the intersection of three triangular planes of soap film.

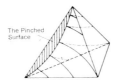

The Pinched
Surface

Fig. 4.13   The result of pinching together two adjacent faces of a tetrahedron.

The angle between adjacent sections of film is 120°. The angle between any two lines can be calculated from this information to be 109°28′. In the analogue solutions to Steiner's problem the angle between any two lines at an intersection was also 120° and there were always three lines meeting at an intersection. This resulted from three planes of soap film, which were perpendicular to the plates, meeting along a line. In equilibrium each soap film plane exerts an equal force at the line, and consequently the angle between any two of the planes of soap film must be 120°. This is a general result for intersecting soap films. In the case of the soap film contained by the tetrahedral framework there are three planar surfaces meeting along any line of soap film and the angles between the surfaces are 120°.

Joseph Plateau[46] discovered experimentally, over a hundred years ago, that soap films contained by a framework always satisfy three geometrical conditions:

1) Three smooth surfaces of a soap film intersect along a line.

2) The angle between any two tangent planes to the intersecting surfaces, at any point along the line of intersection of three surfaces, is 120°.

3) Four of the lines, each formed by the intersection of three surfaces, meet at a point and the angle between any pair of adjacent lines is 109°28′.

These conditions are satisfied by the soap film contained by the tetrahedral framework. The conditions given by Plateau apply to surfaces bounded by any frame. These surfaces do not have to be planar and the lines of soap film need not be straight. It is only recently that Frederick J. Almgren Jr. and Jean E. Taylor[78] have shown that these conditions follow from the mathematical analysis of minimum surfaces and surfaces containing bubbles of air or gas at different pressures, both of which can be described by the Laplace–Young differential equation.

If one of the triangular films formed inside the tetrahedron is broken a new surface will form which has the minimum area contained by the remaining five edges of the tetrahedron (Plate 4.2). The surfaces are no longer all planar. It is seen from Plate 4.2 that the surfaces meet along a curved line. Only the surface with the shape resembling a section of melon is planar. Careful examination will show that any point along the curved line of intersection of the three surfaces has tangent planes that intersect at 120°. If the planar surface is broken, the soap film will take up the minimum area surface contained by the remaining four edges of the tetrahedron. These edges form a skew quadrilateral (Plate 4.3) and so form a surface similar to that in Fig. 4.11.

From the result for the minimum area contained by the six edges of the tetrahedron (Plate 4.1) it might be expected that the minimum surface area contained by the twelve edges of a cubic framework would consist of a series of surfaces each bounded by an edge of the cube and all meeting at a point at the centre. If this did occur it would not be possible for all the surfaces to satisfy Plateau's conditions. The surfaces meeting at the centre of the cube would not intersect at 120° and the angle between two lines would not be 109°28′. What then is the shape of the minimum surface?

A clue to the shape of the minimum surface can be obtained by recalling the solution to the two dimensional 'cube'; the solution to the Steiner problem for four points arranged at the corners of a square (Fig. 3.5). The configuration contains a line, not a point, at the centre. So the cubic framework might be expected to have a square, rather than a line, at the centre.

The minimum surface formed by dipping a cubic framework into a bath of soap solution is shown in Plate 4.4. It contains a 'square' surface at the centre. The 'square' has four curved sides formed by lines of soap film that arise from the intersection of three surfaces. The curved sides of the 'square' intersect at 109°28′, as predicted by Plateau's results. There are also two other lines of film that meet at each corner of the 'square' at the angles predicted by

Plateau. From the edges and corners of the 'square' are surfaces which terminate on the twelve edges of the cube. They are not all planar. The only planar surfaces are those from the vertices of the 'square' to the edges of the framework that are perpendicular to the 'square'. The plane of the 'square' is always parallel to one of the faces of the cube. It can be made to jump from one plane to another by either blowing onto it or by shaking the frame. There are thus three minimum surfaces with the same area. It will be recalled that in the case of the two dimensional problem, with the square array of points, the central line could occur in two positions, parallel to either side of the square.

By rupturing different sections of the minimum surface one can obtain the subset of minimum surfaces associated with a number of the edges of the cube. This number is less than, or equal, to twelve. For example, by breaking the 'square' at the centre with a dry rod, or finger, the minimum surface shown in Fig. 4.14 is obtained. It is a minimum area surface bounded by all twelve edges of the cube. It is not obvious that this surface has a larger, or smaller, surface area than the former surface with a central 'square' and it is difficult to determine the magnitudes of these surface areas.

Fig. 4.14   The minimum surface formed by breaking the 'square' surface at the centre of a cubic framework.

Plate 4.5 shows the result of breaking surfaces linked to two edges of the cube. There are numerous minimum surfaces associated with subsets of edges. They can all be obtained by breaking sections of soap film surface, and provide examples which illustrate Plateau's rules.

The cubic framework produces minimum surfaces with features in common with its two dimensional counterpart, the square array of pins contained by two perspex plates. This is also true of the tetrahedron and the three pins contained by two perspex plates. The octahedral framework also has features in common with its two dimensional counterpart, the hexagon of pins. Consequently it might be expected that there are several minimum surfaces bounded by the twelve edges of the octahedron. This is found to be the case. Dipping an octahedral framework into soap solution can give different minimum surfaces of soap film. These are shown in Plates 4.6(a–e). They can be obtained by withdrawing the framework from the soap solution at different angles. Alternatively one of the minimum surfaces can be produced and then this surface can be perturbed by shaking the framework, or by blowing onto the surface, so that it jumps into another minimum configuration. In this way all the minimum surfaces can be generated.

The most symmetrical surface (Plate 4.6(a)), consists of a number of planar films in the form of triangles and kites. At the centre of the octahedron six kites of soap film meet at a single point. The region surrounding this point is geometrically identical to that at the centre of the minimum surface of the tetrahedron (Plate 4.1). Plates 4.6(b–e) contain a hexagon-like, pentagon-like, square-like, and a quadrilateral-like planar films at the central region of the octahedron. These plane figures have sides that consist of curved lines of film formed by the intersection of three surfaces. The vertices of all 'polygonal' figures have angles of 109°28′.

Plates 4.7 show some of the subset of surfaces formed by breaking one or more of the sections of soap film shown in Plates 4.6. In Plate 4.7(a) the surface forms asymmetrically towards two of the faces of the octahedron. Plate 4.7(b) contains a rhombus-like surface in the plane of one of the squares forming the octahedral framework and Plate 4.7(c) is an attractive surface containing a section of surface in the form of a planar kite.

The angles between adjacent faces of the dodecahedron are 116°36′. As these angles are less than 120° the minimum surface bounded by the edges of a dodecahedral framework will form inside the frame. This surface is shown in Plate 4.8. The surface does not completely enclose a bubble of air because one of the faces of the dodecahedron is open, so that both sides of any section of the soap film surface are at atmospheric pressure. The surface is similar to the two dimensional surface formed by the pentagon of pins contained between two parallel perspex plates (Fig. 3.6). The surface formed by the 5-sided pentagon contains $(5-1)$, four, sides which are linked by surfaces to the edges of the pentagon and thus has one open side. Likewise in the dodecahedron, with 12 faces, the minimum surface contains $(12-1)$, eleven, faces constrained by surfaces joining it to the edges of the dodecahedron.

The dihedral angle of the icosahedron, the angle between adjacent faces, is 138°11′. This is greater than 120° and consequently the minimum surface forms on the faces of the framework. The surface consists of 19 equilateral planar triangular faces and is shown in Plate 4.8(b). The minimum surface, connecting all the edges of the framework, leaves one of the faces open. Thus the pressure difference across any surface is zero and the equilateral triangular sections of soap film are planar.

The investigation of the minimum surfaces formed by any framework can be made using soap films. Having examined the surfaces formed by the Platonic frameworks one is led to investigate the Archimedean[22] frameworks. Every face of an Archimedean framework is a regular polygon, but not all faces contain the same polygon. Every vertex of the frame is congruent to every other vertex. That is, faces are arranged around each vertex in the same order. Surfaces formed by some Archimedean frameworks are shown in Plates 4.9(a–d). The surfaces provide further examples of Plateau's rules.

Another class of simple frameworks are the prisms and the antiprisms. The simplest prism is the triangular prism. If the height of the prism is sufficiently large, the soap film surface formed in a triangular prism will be as shown in Plate 4.10. It will contain three vertical planes of soap film meeting, at angles of 120°, along a vertical line at the central axis of the prism. The sections of the soap film surface at the top and at the bottom of the prism are identical to those in a tetrahedral framework. If the height of the prism is reduced the length of the vertical line formed by the three vertical planes of soap film is reduced. Eventually its length will become zero. At this point the film becomes unstable and will jump into the configuration shown in Fig. 4.15. This surface contains a horizontal triangular-shaped surface. There is a range of prism heights that produce two minimum configurations, one with a small vertical line of soap film along the central axis and the other with a horizontal triangular-shaped surface. The surface can be made to jump from one configuration to the other by perturbing it.

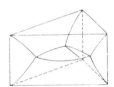

Fig. 4.15   Minimum surface formed by a triangular prism framework with a horizontal 'triangle'.

When the height of the prism is sufficiently great a central horizontal cross-section of the film is similar to that produced by three pins separating two parallel perspex plates which was discussed in Chapter 3. For prisms with bases in the form of polygons one obtains minimum surfaces with a central horizontal section, at the half height of the prism, that are similar to those produced by the polygonal arrangement of pins in the solution of the Steiner problem.

Plate 4.11 shows the minimum surface produced by a helix connected to a central axis. The surface is in the form of a helter-skelter, or corkscrew.

All the surfaces examined in this section have the property that there is no pressure difference across the soap film surface. Thus the Laplace–Young result of section 1.5 gives

$$\frac{1}{R_1} + \frac{1}{R_2} = 0, \tag{4.1}$$

where R_1 and R_2 are the principal radii of the curvature of the soap film surface at any point. In the case of the surface formed inside the tetrahedron the sections of surface are planar, thus R_1 and R_2 are infinite. The result (4.1) is obviously satisfied. In the more general case the radius of curvature R_1 is equal in magnitude and opposite in sign to R_2. This is most simply illustrated in the case of the minimum surface formed by two coaxial rings, the catenoid (Fig. 4.6). It will hold for all the minimum surfaces formed by all frameworks.

The minimum surface area property of soap films has been shown to be a consequence of the minimization of the free energy of the soap films. This minimization condition leads to a differential equation for the surface; the Laplace–Young differential equation for zero excess pressure. In the more general system of soap film surfaces and bubbles the free energy consists of two terms. One contribution from the area of the film, A, and another contribution from the air inside the bubble F_B. This total free energy F is thus given by

$$F = \sigma_f A + F_B, \tag{4.2}$$

where σ_f is the film tension. This total free energy is minimized when the soap film and bubbles reach thermodynamic equilibrium. This minimization condition of the free energy can be re-expressed in terms of the general Laplace–Young differential equation.

4.6   Clusters of Bubbles

Soap bubbles are spherical in shape and consist of a thin shell of water with surfactant molecules at each surface. A bubble encloses some air, or gas, at a pressure that is greater than the external pressure. Clusters of soap bubbles are produced by spherical bubbles that coalesce. Figure 4.16 shows two bubbles that have coalesced. The major part of each bubble consists of a spherical shell of soap film and they are separated by a spherical cap of soap film. In general, clusters of soap bubbles consist of a conglomeration of spherical surfaces and spherical caps of soap film.

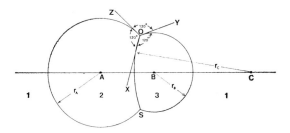

Fig. 4.16 Two coalescing bubbles.

Clusters of soap bubbles provide examples of systems that are more complex than those of simple soap films contained by a fixed boundary. The free energy of the clusters will depend on the surface area of the soap film and the gas contained by the bubbles.

In the Steiner problem, Chapter 3, and the minimum surface area problems, discussed earlier in this chapter, the Laplace–Young equation had a simple form. This resulted from the zero excess pressure across any point on the surface of the soap film. In the case of a bubble, or clusters of bubbles, the excess pressure across any surface is not in general zero. However the Laplace–Young equation can be applied under these more general conditions. Plateau's rules concerning the angles at which surfaces and lines of soap films intersect apply also to the surfaces and lines of soap film produced by clusters of bubbles.

It is worth examining the geometrical properties of small groups of two and three coalescing bubbles in order to relate the geometrical properties to results obtained from the application of the Laplace–Young equation.

4.7 Two Coalescing Bubbles

The excess pressure, p_f, inside a spherical bubble of radius r was shown in section 1.4 to be given by

$$p_f = \frac{2\sigma_f}{r}, \tag{4.3}$$

where σ_f is the film tension. Figure 4.16 shows two coalescing bubbles, centres A and B, of radii r_A and r_B respectively. There are three different pressure regions labelled, 1, 2, and 3. Inside bubble A the pressure is p_2, inside B, p_3, and elsewhere p_1. When the bubbles are in equilibrium these pressures are constant. The surface between region 2 and 1 has an excess pressure across it of p_{21} and a radius of curvature r_A. So from (4.3),

$$p_{21} = \frac{2\sigma_f}{r_A}. \tag{4.4}$$

Similarly, for the spherical surface between regions 3 and 1 with the excess pressure p_{31} and radius of curvature r_B, the excess pressure is given by

$$p_{31} = \frac{2\sigma_f}{r_B}. \tag{4.5}$$

The excess pressure across the surface common to regions 3 and 2, p_{32}, is constant and related to the radius of curvature of the surface, r_C, by

$$p_{32} = \frac{2\sigma_f}{r_C}. \tag{4.6}$$

The centres of the bubbles A and B must, by symmetry, lie on the same straight line as the centre, C, of the sphere of which the common surface, of radius r_C, is part.

The excess pressure p_{21} must be equal to the sum of the excess pressures between region 2 and region 3, p_{23}, and p_{31}. So

$$p_{21} = p_{23} + p_{31}. \tag{4.7}$$

Substituting (4.4), (4.5), and (4.6) into (4.7),

$$\frac{2\sigma_f}{r_A} = -\frac{2\sigma_f}{r_C} + \frac{2\sigma_f}{r_B}, \tag{4.8}$$

as $p_{23} = -p_{32}$. Hence

$$\frac{1}{r_B} = \frac{1}{r_A} + \frac{1}{r_C}. \tag{4.9}$$

There are many examples in science of reciprocal relationships of this kind. For example the focal length of a thin lens is related to the object and image distances, measured from the lens, by such a relationship. Also the total resistance due to two resistors in parallel is obtained by such a relationship, likewise for the sum of two series capacitors. The two coalescing bubbles could, in principle, be used as an analogue system for solving lens problems, resistor problems, etc. In the case of the lens problem it is necessary to produce two coalescing bubbles, one with its radius proportional to the object distance and the other proportional to the focal length of the lens, in order to calculate the image distance which is proportional to the radius of the common surface. However, in practice, it is difficult to set up such an analogue system.

The reciprocal relation has been obtained by using the Laplace–Young equation for each spherical surface. In this example it is possible to show that the general property concerning the intersection of three surfaces of soap film at angles of 120 ° provides an alternative method of obtaining this reciprocal relation. This indicates that the Laplace–Young equation and the geometrical properties of soap films are closely related. Now let us examine these geometrical properties in order to obtain an alternative derivation of the reciprocal relationship.

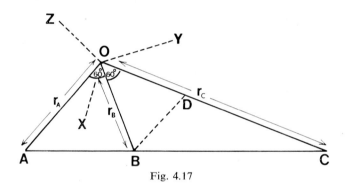

Fig. 4.17

Figure 4.17 shows the upper half of a section through the line joining the centres of curvature of the three external surfaces of the bubbles in Fig. 4.16. These centres are joined to O, the point of intersection of the three surfaces in a plane through AC. The tangent planes at O, OZ, OY, and OX intersect at angles of 120 °. The radii of curvature OA and OC (Fig. 4.17) are respectively perpendicular to these tangent planes, OZ and OX. So they must intersect at 120 °, that is

$$A\hat{O}C = 120°. \qquad (4.10)$$

As the radii of the surfaces at O are perpendicular to the tangent planes at O,

$$A\hat{O}Z = 90°, \qquad (4.11)$$

and

$$B\hat{O}Y = 90°. \qquad (4.12)$$

In Fig. 4.17,

$$A\hat{O}B = X\hat{O}Z + X\hat{O}Y - A\hat{O}Z - B\hat{O}Y. \tag{4.13}$$

Using the geometrical property that tangent planes intersect at $120°$, and the results (4.11) and (4.12),

$$A\hat{O}B = 120° + 120° - 90° - 90°, \tag{4.14}$$

$$= 60°. \tag{4.15}$$

Thus from (4.10) and (4.15),

$$B\hat{O}C = 60°. \tag{4.16}$$

In order to obtain the reciprocal relation it is necessary to construct a line through B parallel to OA (Fig. 4.17) meeting OC at D. The triangles AOC and BDC are similar. Thus the ratio of the sides have the property that

$$\frac{BD}{AO} = \frac{DC}{OC}, \tag{4.17}$$

or

$$DC = \frac{BD}{AO} OC, \tag{4.18}$$

$$= \frac{BD}{r_A} r_C. \tag{4.19}$$

As OA is parallel to DB,

$$O\hat{B}D = A\hat{O}B, \tag{4.20}$$

from (4.15),

$$= 60°. \tag{4.21}$$

In triangle BOD the angles add up to $180°$ to give, from (4.16) and (4.21),

$$O\hat{D}B = 60°. \tag{4.22}$$

Therefore triangle OBD is equilateral and

$$OD = BD = r_B. \tag{4.23}$$

From (4.19) and (4.23),

$$DC = \frac{r_B}{r_A} r_C. \tag{4.24}$$

Now

$$OD = OC - DC, \tag{4.25}$$

hence from (4.23) and (4.24),

$$r_B = r_C - \frac{r_B}{r_A} r_C. \tag{4.26}$$

Dividing by $r_B r_C$,

$$\frac{1}{r_B} = \frac{1}{r_A} + \frac{1}{r_C}. \tag{4.27}$$

Thus the reciprocal relation (4.27) has been obtained by purely geometrical methods using the geometrical properties of intersecting surfaces.

4.8 Three Coalescing Bubbles

In the case of two coalescing bubbles the radii of curvature of the three spherical surfaces, r_A, r_B, and r_C, are related by the reciprocal relation

$$\frac{1}{r_B} = \frac{1}{r_A} + \frac{1}{r_C}. \tag{4.28}$$

The centres of the spheres of which the surfaces form part lie on a straight line. For three coalescing bubbles (Fig. 4.18) the reciprocal relation holds for each pair of bubbles. Thus in addition to (4.28) there are reciprocal relations for the radii of the surfaces along the directions AD and BD, where D is the centre of the external surface of the small bubble.

If E is the centre of curvature and r_E the radius of curvature of the surface separating the bubbles with centres A and D, the reciprocal relation gives,

$$\frac{1}{r_D} = \frac{1}{r_A} + \frac{1}{r_E}. \tag{4.29}$$

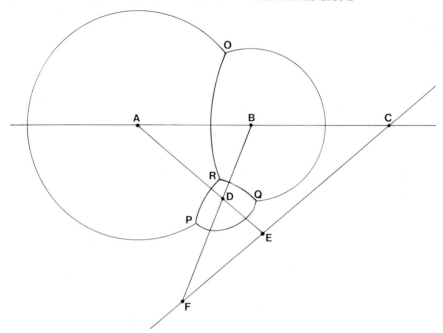

Fig. 4.18   Three coalescing bubbles.

Similarly for bubbles with centres B and D, assuming the intermediate sur
face has radius r_F and centre F along BD,

$$\frac{1}{r_D} = \frac{1}{r_B} + \frac{1}{r_F}. \tag{4.30}$$

The centres of the intermediate surfaces C, E and F lie respectively on the
extensions of AB, AD and BD. It is possible to prove that C, E and F all lie
on a straight line. Let us prove this result in the subsequent part of the section.

There is a geometrical theorem due to Menelaus[19] that states that the condition that the points C, E and F lie on the same straight line is

$$\left(\frac{BF}{FD}\right)\left(\frac{DE}{EA}\right)\left(\frac{AC}{CB}\right) = -1, \tag{4.31}$$

where the sign convention takes into account the direction of the lines. We
shall use this result to show that C, E and F lie on a straight line.

Using Fig. 4.17 we can obtain a geometric result for two coalescing
bubbles that will enable us to use Eq. (4.31). The similar triangles AOC and
BDC give

$$\frac{AC}{CB} = -\frac{AO}{BD}. \tag{4.32}$$

As triangle OBD is equilateral, $BD = OB = r_B$ and hence

$$\frac{AO}{BD} = + \frac{r_A}{r_B}. \tag{4.33}$$

Therefore from (4.32) and (4.33),

$$\frac{AC}{CB} = -\frac{r_A}{r_B}. \tag{4.34}$$

This result applies to any two bubbles in Fig. 4.18. So in addition to result (4.34) for Fig. 4.18, we can obtain,

$$\frac{BF}{FD} = -\frac{r_B}{r_D}, \tag{4.35}$$

and

$$\frac{DE}{EA} = -\frac{r_D}{r_A}. \tag{4.36}$$

Hence from (4.34), (4.35) and (4.36),

$$\left(\frac{BF}{FD}\right)\left(\frac{DE}{EA}\right)\left(\frac{AC}{CB}\right) = \left(-\frac{r_B}{r_D}\right)\left(-\frac{r_D}{r_A}\right)\left(-\frac{r_A}{r_B}\right), \tag{4.37}$$

$$= -1. \tag{4.38}$$

The points C, E and F satisfy Menelaus's theorem and thus C, E and F all lie on a straight line.

4.9 Bubbles, Bubble Rafts, and Foam

Clusters of bubbles illustrate Plateau's three rules. In the two cases studied so far the points, such as O and S in Fig. 4.16 and P, Q, R and O in Fig. 4.18, are formed from the intersection of three surfaces. The angles of intersection of the tangent planes to these surfaces are $120°$.

When a cluster is formed from bubbles of different radii the internal surfaces will be curved. This is due to the pressure difference between neighbouring bubbles and is illustrated by the internal surfaces in Figs. 4.16 and 4.18. However if the bubbles were originally of the same radius this pressure difference is zero. Thus the common internal surface will have zero curvature and will have the shape of a disc. A cluster of coalescing bubbles formed by bubbles of equal radius that are all symmetrical about a plane, such as those shown in Fig. 4.19, have internal surfaces represented by straight lines in the symmetrical sectional diagram. The arrangement of the lines is characterized by the $120°$ angles between the straight lines. The different configurations of

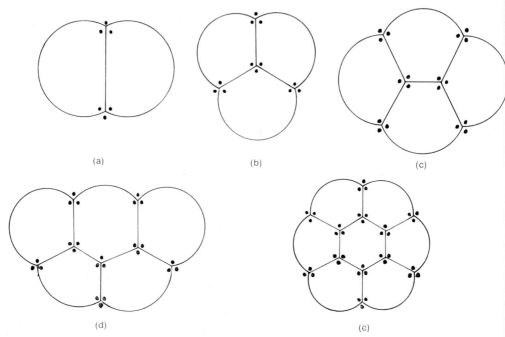

(a)   (b)   (c)

(d)   (e)

Fig. 4.19   Groups of bubbles with equal radii, the 120° angles are indicated by dots.

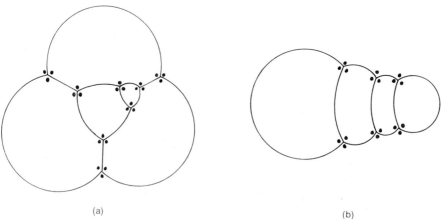

(a)   (b)

Fig. 4.20   Groups of bubbles with unequal radii, the 120° angles are indicated by dots.

these lines are identical to those produced using pins and parallel plates in the solution to the Steiner problem. Figure 4.20 shows symmetrical sections through clusters of bubbles, of different radii, that are symmetrical about a plane. The 120° angles are indicated by dots.

Bubbles can produce hemispherical clusters when formed on a wet surface or on the surface of a bath of soap solution. Figure 4.21 shows symmetrical hemispherical clusters of bubbles that are symmetrical about a plane. A symmetrical vertical section has been taken through the bubbles and the 120° angles of intersection of the surfaces have been indicated by dots.

If a large number of hemispherical bubbles of equal size are produced on the surface of a bath of soap solution, using a uniform pressure jet of air, they will tend to form a plane hexagonal lattice, or honeycomb lattice, of bubbles (Plate 4.12(a)). Lawrence Bragg[81] first used such rafts of bubbles to show that two dimensional lattices of bubbles have many features in common with three dimensional atomic crystalline lattices. These rafts have become known as Bragg rafts.

Dislocations can be produced in the Bragg raft. This is most easily carried out by rupturing consecutive bubbles in one half of the bubbles in a line of bubbles passing through the central region of the bubble raft. Thus one of the lattice lines of bubbles is terminated in the central region of the raft producing a dislocation (Plate 4.12(b)). The dislocation is easily observed if one views the Plate along the appropriate lattice direction in the plane of the Plate.

A defect can be created in the bubble lattice by increasing, or decreasing, the size of one of the bubbles (Plate 4.12(c)). An interstitial bubble, also, is easily introduced into the Bragg raft.

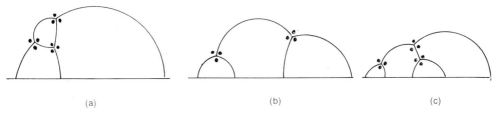

(a)   (b)   (c)

Fig. 4.21   Groups of hemispherical bubbles, the 120° angles are indicated by dots.

The lattice ordering in a perfect lattice raft can be partially destroyed by stirring it with a rake. This will produce a number of small ordered regions of bubbles which have the original regular periodic structure. These regions are separated from one another by grain boundaries. These are boundaries where the lattice ordering has been destroyed (Plate 4.12(d)). The ordered regions tend to grow if the raft is left undisturbed, and the total length of the grain boundaries decreases. This produces 'recrystallization' of the bubble raft.

All of these two dimensional bubble raft demonstrations simulate the arrangement of atoms in crystalline materials. Dislocations, lattice defects, grain boundaries and recrystallization are all phenomena that occur in three dimensional crystalline materials.

The similarity between the behaviour of hemispherical bubbles in a bubble raft and atoms in a crystalline lattice is not so surprising when one considers the forces acting between two hemispherical bubbles on a Bragg raft. It is found that, when the centres of two bubbles are separated by a distance that is large compared with the diameter of a bubble, the bubbles attract each other weakly. As the bubble–bubble separation decreases the attractive force increases and reaches a maximum value when the centres are separated by a distance just greater than the diameter of a bubble. As the separation is decreased further the force, which is attractive, rapidly decreases and becomes strongly repulsive. The repulsive nature of the force is due to the coalescence of the bubbles. The fixed amount of air in each bubble produces an increasingly strong repulsive force as the separation of the bubbles is decreased. The area of the common surface separating the bubbles increases and consequently the repulsive force, which is the product of the air pressure and the area of the common surface, increases.

This force which is strongly repulsive at short distances and weakly attractive at large distances, with the maximum attractive force at a separation just greater than the diameter of a bubble, resembles the Lennard–Jones force between atoms. Consequently it is to be expected that the bubbles will have a similar behaviour to a two dimensional lattice of interacting atoms, and have many characteristics in common with three dimensional atomic crystalline structures. Mr. M. M. Nicholson[107] has worked out, in detail, the forces between two bubbles in a Bragg raft using the Laplace–Young equation to describe the surface configuration of the two interacting hemispherical bubbles.

Fig. 4.22   Two cylindrical bubbles in contact.

Bubbles formed between horizontal plates which are separated by a constant vertical height and touch both plates will, by symmetry, be perpendicular to the plates and have vertical sides. In the case of a single bubble, it will be cylindrical in shape. The ends of the cylinder being formed by the parallel plates. Groups of bubbles will likewise have vertical sides. These vertical sides will be planar if the pressure difference across the surface is zero. This occurs in the case of the common surface formed by the coalescence of cylindrical bubbles of equal size. Figure 4.22 shows two cylindrical bubbles of equal size with a common planar surface. The Plateau rules will be satisfied by the surfaces and lines of soap film in these bubbles.

Foam is an agglomeration of a large number of different bubbles (Plate 4.13). Each bubble in the foam is a polyhedral cell with a number of different faces. Each face is curved as a result of the excess pressure across it. If the pressure in two adjacent cells is the same, the excess pressure across the surface is zero and the separating surface is planar. The foam will contain, only, three film surfaces intersecting along lines at $120°$ and four lines of soap film meeting at a point with adjacent lines intersecting at $109°28'$.

It is possible to obtain an approximation for the average number of faces, vertices and edges associated with a polyhedral cell of foam[115] by considering the average cell in the foam as a *regular* polyhedron with planar faces. This is an approximation as the faces are known to be curved. However, as the curvatures of the faces are small it is a good approximation. Formulae valid for polyhedra, with planar faces, can then be applied using the result that the angles between the edges of a polyhedral cell of foam are $109°28'$.

If each face of the *regular* polyhedron, which is an approximation to the average polyhedral cell, has n sides and the angles at the corners of the faces are assumed to be θ, then the sum of the angles of the n-edged polygon is

$$n\theta = 180n - 360. \tag{4.39}$$

Extending this result to the average polyhedral cell in which n is not an integer and substituting $\theta = 109.47°$ gives

$$n = 5.10. \tag{4.40}$$

So, from this calculation, the average number of edges associated with each face is 5.10.

It is possible to relate the sum of the face angles surrounding any vertex of the regular polyhedron, ϕ, to the number of vertices, v. Summing over all internal angles of all the, f, faces of the polyhedron,

$$v\phi = fn\theta, \tag{4.41}$$

as each polygonal face has n internal angles. From (4.39), (4.41) becomes

$$v\phi = f(180n - 360). \tag{4.42}$$

Therefore

$$v\phi = 180fn - 360f. \tag{4.43}$$

The total number of the edges, e, of the polyhedron is half the product of the number of faces and the number of edges per face, so

$$fn = 2e. \tag{4.44}$$

Substituting (4.44) into (4.43),

$$v\phi = 360(e - f). \tag{4.45}$$

Euler's theorem[14] relating e, f and v states that

$$e - f = v - 2. \tag{4.46}$$

Substituting (4.46) into (4.45),

$$v\phi = 360(v - 2). \tag{4.47}$$

Therefore,

$$v = \frac{720}{(360 - \phi)}. \tag{4.48}$$

Now applying this result to the average polyhedral cell of foam, with $\phi = 3\theta = 3(109.47\,°)$,

$$v = 22.79. \tag{4.49}$$

Finally, using Euler's theorem, (4.46), and (4.44),

$$v - \tfrac{1}{2}fn + f = 2. \tag{4.50}$$

Thus

$$f = \frac{2(v - 2)}{(n - 2)}. \tag{4.51}$$

Substituting from (4.40) and (4.49) gives,

$$f = 13.39. \tag{4.52}$$

Thus this approximation gives the average number of faces of a foam cell as 13.39, the average number of vertices as 22.79, and the average number of sides to each face as 5.10.

4.10   Anti-Bubbles

When soap solution is poured gently from a beaker, held just above the surface of a bath containing soap solution, onto the soap solution in the bath, or is dropped from a burette, individual drops of soap solution often remain in the surface and move in the surface just like ball bearings [116] moving on a flat surface. They move away radially from the point of impact. As these drops are repelled by the walls of the container and are attracted by each other, they tend to form groups and coalesce into larger drops. These droplets are unstable and have a lifetime of about 10 seconds. After this time they coalesce suddenly with the bulk of the soap solution.

If the soap solution is poured rapidly into the bulk solution, it will sometimes penetrate deeper into the bulk solution and form an isolated 'bag' of soap solution surrounded by a thin skin of air (Fig. 4.23). It is not always possible to obtain this result. However with experience 'bags' of soap solution can be produced with up to 20 mm diameter. A 'bag' is a spherical globule of soap solution surrounded by a thin shell of air existing in the bulk of the soap solution. This is the anti-bubble.[116],[121] It is the inverse of a soap bubble. The soap film has been replaced by an air film, and the air contained inside and outside the soap bubble is replaced by soap solution.

FIG 4.23

Fig. 4.23   An anti-bubble.

Once an anti-bubble has been formed it will move, under its own momentum, down into the bulk of the fluid. The buoyancy of the surrounding air will provide an upthrust which will cause it to slow down and eventually rise to the surface. On reaching the surface it will either bounce back into the fluid or come to rest under the surface as a sphere or a hemisphere.

Interference colours can be observed from the shell of air which is approximately 0.001 mm thick. The lifetime of these anti-bubbles may range from seconds to several minutes. They disintegrate suddenly, like the fluid droplets, and produce numerous air bubbles.

The anti-bubbles are intrinsically unstable. The shell of air is contained between two spherical surfaces of fluid, one being convex and the other concave. The saturated vapour pressure of the inner surface is larger than the outer one. This will lead to evaporation from the inner sphere of fluid and condensation on the other concave surface of fluid. Also the film of air is intrinsically unstable against 'draining'.

4.11   Bubbles Trapped in Frames

It is possible to trap a soap bubble inside a framework so that it is constrained by surfaces attached to the framework. For example, in the case of a tetrahedral framework one first forms the minimum surface shown in Plate 4.1. The tetrahedron is then partially resubmerged in the bath of soap solution so that an air bubble is trapped between the minimum surface and the surface of the soap solution in the bath. When the framework is withdrawn from the soap solution the bubble of air will be trapped in the central region of the framework. It will have the symmetry of the framework and will be constrained in the form of a tetrahedral bubble by surfaces attached to the edges of the framework and the bubble. It is shown in Plate 4.14(a).

Plates 4.14(b), (c), (d) and (e) show bubbles formed inside frameworks with the geometry of a cube, octahedron, triangular prism and dodecahedron respectively; (b), (d) and (e) contain bubbles with the symmetry of the containing framework. The size of the bubbles depends on the quantity of air trapped in the framework. If a bubble is ruptured the air will be released and it will reform a minimum surface associated with the framework. Alternatively if the surfaces constraining the bubble are systematically broken, the bubble will become increasingly spherical. Finally when all the constraining surfaces are broken the bubble will be spherical and free of the framework.

The system consisting of a soap film plus air trapped in a bubble has a free energy, F, at constant temperature and external pressure that consists of two contributions. There is a contribution, $\sigma_f A$, from the surface area A, where σ_f is the film tension of the soap film. There is in addition a contribution from the air trapped in the bubble, F_B. F is the sum of the two contributions, $\sigma_f A$ and F_B, (Eq. (4.2)). F is minimized to give the equilibrium configurations shown in Plates 4.14(a–h). Plates 4.14(f–h) are three Archimedean frames containing bubbles.

All the bubbles formed inside frameworks satisfy Plateau's rules. Three surfaces meet along a line at angles of $120°$ and four lines intersect at a point so that any two lines meet at $109°28'$. Bubbles can be produced in any framework by the same procedure.

The frameworks may be constructed from any easily wetted materials; wood, metals, and many plastics (not teflon). The soap solution can be prepared with water plus approximately one per cent of any washing-up liquid. This solution will produce films and bubbles that last for up to 20 seconds. For longer lasting films and bubbles a special soap solution should be prepared according to one of the prescriptions given in section 1.7. The demonstrations can be performed on a large scale using 20 gallons of water contained in a dustbin plus one per cent of washing-up liquid. The dimensions of the frameworks can then be of the order of 20 cm.

The surfaces and bubbles may be projected onto a large screen by shadow projection using a strong point source of light such as that produced by a quartz-iodine (halogen) lamp.

It is worth mentioning an unusual and interesting demonstration in which a soap bubble is repeatedly bounced off of a disc of soap film as it always attracts considerable attention. It requires the production of a large soap bubble, about 20 cm in diameter, and a ring of wire, about 30 cm in diameter. The bubble can be bounced many times from the disc of soap film in the ring without coalescing with it, even when the soap solution is prepared from household washing-up liquid and water. The simplest method of performing the demonstration uses a disc of soap film contained in a ring of wire. The horizontal disc of soap film can be oscillated up and down by moving the ring up and down. A bubble can then be formed by moving the ring rapidly in a horizontal direction so that a volume of air is totally enclosed. The bubble will break away from the frame and leave a disc of soap film in the ring. The bubble can now be bounced from the film remaining in the ring (Plates 4.15(a) and (b)).

It would be most convenient if it were possible to form these surfaces and bubbles from a plastic solution that would rapidly solidify once the film had formed inside the large frameworks. Unfortunately there does not appear to be a suitable plastic solution on the market that is capable of producing large stable films. However there are some plastic solutions that will form films and bubbles inside frameworks providing the dimensions of the frames are sufficiently small. Frames with dimensions of about 2 cm are usually satisfactory. The plastic solution, that has been used successfully, is marketed with a child's toy that enables one to produce imitation plastic flowers by dipping loops of wire into a plastic solution. A plastic film forms in the loop, after withdrawal from the solution, and rapidly solidifies.

4.12 Radiolarians

D'Arcy Wentworth Thompson[37] has pointed out, in his book *On Growth and Form*, that there are microscopic marine organisms called Radiolarians whose skeletons have the same basic shape as soap bubbles trapped in frameworks. The living organisms consist of a mass of protoplasm surrounded by a froth of cells. The silica bearing fluid of the organism accumulates along the intersection of foam-like surfaces and forms a skeleton.

When the organism dies the skeleton falls to the bottom of the ocean. Many of these skeletons have shapes similar to bubbles trapped in frameworks. Some of the Radiolarian skeletons are shown in Fig. 4.24. These skeletons were obtained by the biologist Ernst Haeckel on the Challenger Expedition of 1873–76. He found the skeletons in samples of mud taken from the ocean bed. Many of the samples are to be found in the Natural History Museum, London, England. Some superb glass models of these skeletons are to be seen at The Natural History Museum in New York City.

The Nassellarian skeleton, *Callimitra agnesae*, is shown in Fig. 4.24(a). It resembles the bubble and surfaces formed in a tetrahedral framework (Plate 4.14(a)). *Lithocubus geometricus*, the skeleton shown in Fig. 4.24(b), is similar to that of Plate 4.14(b), the bubble inside the cubic framework. *Prismatium tripodium* (Fig. 4.24(c)) is similar to the bubble inside the triangular prism framework (Plate 4.14(d)). Other interesting Radiolarian skeletons are shown in Fig. 4.24(d–g).

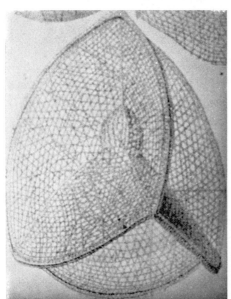

Fig. 4.24 (a)
Callimitra agnesae

Fig. 4.24 (b)   Lithocubus geometricus

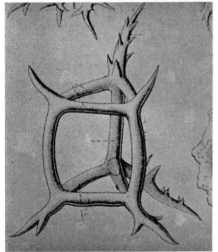

Fig. 4.24 (c)   Prismatium tripodium

Fig. 4.24 (d)   Archicircus rhombus

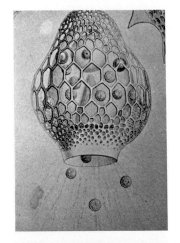

Fig. 4.24 (e)　Cyrtocalpis urceolus

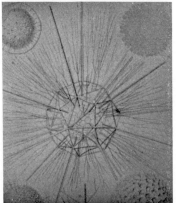

Fig. 4.24 (f)　Drymosphoera
　　　　　　　　dendrophora

Fig. 4.24 (g)　Auloscena mirabilis

5 THE LAPLACE–YOUNG EQUATION

5.1 Introduction

In section 1.5 we discussed the equilibrium of a spherical droplet and a spherical bubble. The excess pressure, p, across the surface of a droplet of radius R and surface tension σ was shown to be,

$$p = \frac{2\sigma}{R}. \tag{5.1}$$

For a soap bubble,

$$p_f = \frac{2\sigma_f}{R}, \tag{5.2}$$

where σ_f is the film tension and p_f is the excess pressure across the soap film. The concept of a film tension, σ_f, was introduced as the surface tension associated with each surface of a soap film may differ from that produced by the surface of soap solution in contact with its bulk fluid. It was pointed out that these results can be generalized to obtain the excess pressure, p, at any point across a surface formed by the interface of two fluids that do not mix, or immiscible fluids. This includes the case of two different liquids and a liquid in contact with a gas or vapour. The general result is usually written in terms of the two principal radii of curvature, R_1 and R_2, at any point on the surface. The principal radii at any point are the maximum and minimum radii of curvature at the point on the surface. The general result for a surface with surface tension, or interfacial tension, σ, is

$$p = \sigma \left(\frac{1}{R_1} + \frac{1}{R_2} \right) \tag{5.3}$$

This is the Laplace–Young equation. It reduces to Eq. (5.1) in the case of a spherical droplet of radius R as $R_1 = R_2 = R$. Equation (5.2) is obtained for a soap bubble, with two surfaces, in which $\sigma_f = 2\sigma$.

The Laplace–Young equation was used, in section 4.4, to discuss the solution to such problems as the minimum surface area contained by two rings and that contained by a skew quadrilateral. In these problems the soap film produces a double surface. The pressure difference across the soap film is zero as the pressure is the same on both sides of the film. The Laplace–Young equation, (5.3), reduces to

$$\frac{1}{R_1} + \frac{1}{R_2} = 0. \tag{5.4}$$

This result is valid for surfaces contained by any boundary providing no bubbles are present and applies, for example, to the soap films contained by the frameworks examined in section 4.5. It can be shown analytically that all minimum surfaces contained by a fixed boundary must satisfy Eq. (5.4). In Appendix IV it is proved, for a restricted set of boundaries, that the minimum area condition leads to Eq. (5.4).

In the most general problem, where bubbles are present and the pressure difference across the soap film is non zero, the minimization of the total free energy leads to the general form of the Laplace–Young equation, (5.3).

In Cartesian coordinates the surface forming the boundary between two immiscible fluids, or a fluid and its vapour, can be expressed as

$$z = f(x, y), \tag{5.5}$$

where $f(x,y)$ is a function of coordinates x and y. It is convenient to introduce the mean curvature of the surface, H, given by

$$H = \frac{1}{2}\left(\frac{1}{R_1} + \frac{1}{R_2}\right). \tag{5.6}$$

The Laplace–Young equation, (5.3), can be re-expressed as a differential equation[20] in Cartesian coordinates using the Cartesian differential form of (5.6) and (5.5), as

$$H = \frac{(1+f_y^2)f_{xx} - 2f_x f_y f_{xy} + (1+f_x^2)f_{yy}}{2(1+f_x^2+f_y^2)^{3/2}}, \tag{5.7}$$

where

$$f_x = \frac{\partial f}{\partial x}, \; f_y = \frac{\partial f}{\partial y}, \; f_{xx} = \frac{\partial^2 f}{\partial x^2}, \; f_{yy} = \frac{\partial^2 f}{\partial y^2}, \; \text{and} \; f_{xy} = \frac{\partial^2 f}{\partial x \partial y}. \tag{5.8}$$

The Laplace–Young equation, (5.3), thus becomes

$$\frac{p}{\sigma} = \frac{(1+f_y^2)f_{xx} - 2f_xf_yf_{xy} + (1+f_x^2)f_{yy}}{(1+f_x^2+f_y^2)^{3/2}}.$$ (5.9)

This is a non-linear, second order, partial differential equation. The different characteristic surfaces are shown in Fig. 5.1 together with the corresponding sign of H. The arrow in the figure indicates the direction of the pressure difference, p.

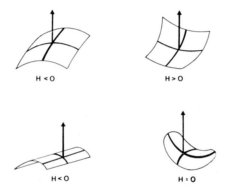

H < 0   H > 0

H < 0   H = 0

Fig. 5.1   The mean curvature, H, for different surface configurations.

For the problems in which $p = 0$ the differential Eq. (5.9) reduces to

$$(1+f_y^2)f_{xx} - 2f_xf_yf_{xy} + (1+f_x^2)f_{yy} = 0.$$ (5.10)

The solution of this equation, with appropriate boundary conditions, will give the minimum area contained by a closed boundary. Such problems as the minimum area contained by two parallel coaxial rings and the minimum area contained by a skew quadrilateral can be solved using this equation. In principle all the minimum area problems discussed in Chapter 4 can be solved using (5.10). However most of them have not been found to be amenable to an analytic solution.

The differential equation, Eq. (5.9), can be applied to the solution of such problems as: the equilibrium configuration of a liquid drop (Fig. 5.2(a)); a pendant drop (Fig. 5.2(b)); and the shape of the meniscus formed in a capillary tube (Fig. 5.2(c)). All of these problems have an excess pressure across the boundary surface of the fluid that is non zero. The excess pressure in these examples is due to the gravitational field. For example in Fig. 5.2(c) this is produced by the height of fluid above the general level of the fluid in the bath. So the excess pressure, p, will typically be of the form,

$$p = az + b,   \tag{5.11}$$

where a and b are constants and z is the vertical height of the fluid. The solution to these problems will be examined in this chapter. However before attempting to solve them a derivation of the Laplace–Young equation will be given.

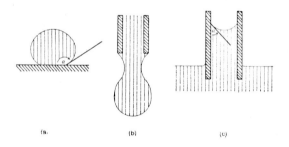

Fig. 5.2   (a) A liquid drop. (b) A pendant drop. (c) Rise of liquid in a capillary tube.

5.2   Derivation of the Laplace–Young Equation

Consider a small curvilinear rectangular element of the surface separating two fluids, $ABCD$, of area S (Fig. 5.3). The sides of the rectangle (Fig. 5.3) have lengths x and y. The radii of curvature of the sides of the rectangle are r_1 and r_2, as indicated in Fig. 5.3. The radii of curvature at A and B in the plane through the x-axis meet at O_1, and the radii of curvature at B and C in the plane through the y-axis meet at O_2.

Now let the surface element $ABCD$ undergo a virtual displacement, δu, normal to the surface under the action of the excess pressure, p. The new position of $ABCD$ is $A'B'C'D'$. The radii of curvature will increase to $r_1 + \delta u$ and $r_2 + \delta u$, and the sides of the surface element will increase to $x + \delta x$ and $y + \delta y$ as indicated in Fig. 5.3. If this virtual displacement is performed under isothermal conditions, the work done by the excess pressure,

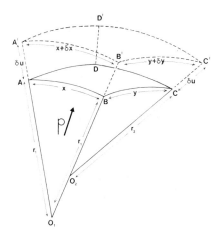

Fig. 5.3 A curvilinear rectangular element of a surface separating two fluids.

δW, will be equal to the increase in the surface energy, δF, due to the change in surface area of the element, δS. That is,

$$\delta W = pS\delta u. \tag{5.12}$$

The increase in energy on the surface

$$\delta F = \sigma \delta S, \tag{5.13}$$

where σ is the interfacial tension, or surface tension, between the two fluids. Expressing (5.13) in terms of x and y,

$$\delta F = \sigma[(x + \delta x)(y + \delta y) - xy]. \tag{5.14}$$

The segments $O_1A'B'$ and O_1AB are similar, hence

$$\frac{x+\delta x}{r_1+\delta u} = \frac{x}{r_1}. \tag{5.15}$$

Thus

$$x+\delta x = x\left(1 + \frac{\delta u}{r_1}\right). \tag{5.16}$$

Similarly for segments $O_2B'C'$ and O_2BC,

$$y+\delta y = y\left(1 + \frac{\delta u}{r_2}\right). \tag{5.17}$$

Substituting from (5.16) and (5.17) into (5.14),

$$\delta F = \sigma\left[xy\left(1 + \frac{\delta u}{r_1}\right)\left(1 + \frac{\delta u}{r_2}\right) - xy\right] \tag{5.18}$$

$$= \sigma xy\delta u\left(\frac{1}{r_1} + \frac{1}{r_2}\right) + O(\delta u^2), \tag{5.19}$$

$$= \sigma S\delta u\left(\frac{1}{r_1} + \frac{1}{r_2}\right) + O(\delta u^2), \tag{5.20}$$

as $S = xy$.

For small displacements, δu, terms of $O(\delta u^2)$ in (5.20) may be neglected. Equating δW and δF, from (5.12) and (5.20),

$$pS\delta u = \sigma S\delta u\left(\frac{1}{r_1} + \frac{1}{r_2}\right). \tag{5.21}$$

Therefore

$$p = \sigma\left(\frac{1}{r_1} + \frac{1}{r_2}\right). \tag{5.22}$$

This is valid for any orthogonal curvilinear system of coordinates in which x and y are measured in the interfacial surface. If we choose the coordinate system which forms the principal, maximum and minimum, radii of curvature R_1 and R_2 at any point in the surface, then (5.22) becomes

$$p = \sigma \left(\frac{1}{R_1} + \frac{1}{R_2} \right). \tag{5.23}$$

This is the well known Laplace–Young equation. If a radius of curvature R_1, or R_2, is in the opposite direction to that shown in Fig. 5.3 it will have a negative sign. That is, a radius of curvature is positive if it is convex in the direction of the pressure and negative if it is concave in this direction.

A soap film has two parallel surfaces. Thus applying the Laplace–Young equation to each surface the excess pressure, p_f, across the film is

$$p_f = \sigma_f \left(\frac{1}{R_1} + \frac{1}{R_2} \right), \tag{5.24}$$

where $\sigma_f = 2\sigma$. The weight of the soap film element is negligible compared with the surface tension forces acting on the element. In thermodynamic equilibrium its thickness will typically be in the range 50–300 Å. Consequently the weight of the film may be neglected. This is also true of the interfacial layer between two immiscible fluids.

The Laplace–Young equation determines the shape of the interfacial surface providing the excess pressure, p, is known at all points on the surface. We have emphasised that it is not possible, in general, to solve Eq. (5.9). Many problems have symmetry properties which enable the Laplace–Young equation to be simplified, and may lead to an analytic solution.

5.3 A Liquid Surface in Contact with an Inclined Plane

The vertical cross-section of a surface of a liquid in contact with an inclined plate is shown in Fig. 5.4. All cross-sectional planes perpendicular to the plate will be identical by symmetry. At large distances from the plate the liquid surface will be horizontal. Let the angle of contact of the liquid with the plate be α and the angle of inclination of the plate with the horizontal be β. The density of the gas, or vapour, above the surface is ρ_1 and that of the liquid ρ_2. It will be convenient to introduce a density ρ defined by

$$\rho = \rho_2 - \rho_1. \tag{5.25}$$

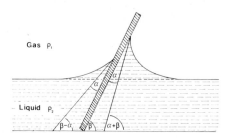

Fig. 5.4   Cross-section of a liquid surface in contact with an inclined plate.

The Laplace–Young equation for the surface of the liquid is

$$p = \sigma\left(\frac{1}{R_1} + \frac{1}{R_2}\right). \qquad (5.26)$$

If R_1 is the principal radius of curvature at a point on the surface of the liquid, shown in the sectional diagram Fig. 5.5 or 5.4, then the principal radius of curvature, R_2, perpendicular to this plane is infinite. So (5.26) becomes

$$p = \sigma\left(\frac{1}{R_1}\right). \qquad (5.27)$$

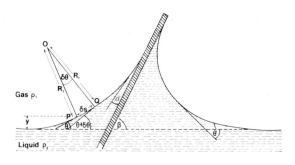

Fig. 5.5   An element of surface PQ.

A small element of the surface of the liquid, such as PQ in Fig. 5.5, has a pressure just above it that is equal to the pressure in the horizontal surface of the liquid at infinity. This is equal to the pressure at any point in the horizontal plane of liquid tangential to the surface at infinity. This is indicated by the broken line in Fig. 5.5. Consequently the pressure just below the surface element PQ is equal to the external atmospheric pressure minus the pressure due to the height, y, of liquid between PQ and the horizontal plane indicated by the broken line in Fig. 5.5. Thus the pressure difference across the surface is,

$$p = \rho g y. \tag{5.28}$$

The Laplace–Young equation (5.27) now becomes,

$$\rho g y = \frac{\sigma}{R_1}, \tag{5.29}$$

where y is measured from the horizontal surface indicated by the broken line in Fig. 5.5.

In order to simplify Eq. (5.29) it is necessary to express R_1 in terms of a more convenient set of coordinates. It will simplify the analysis to construct tangents and normals to the surface at P and Q. Let the tangents meet the horizontal at angles θ and $\theta + \delta\theta$ respectively. The normals will intersect at the point O (Fig. 5.5) with angle $\delta\theta$ and have lengths R_1. If the length of the arc PQ is δs, then

$$R_1 = \frac{\delta s}{\delta\theta}. \tag{5.30}$$

If δy is the vertical height of Q above P, then resolving δs vertically,

$$\delta y = \delta s \sin\theta. \tag{5.31}$$

Substituting for R_1, from (5.30), into (5.29),

$$\rho g y = \sigma \frac{d\theta}{ds}. \tag{5.32}$$

Substituting for ds from (5.31),

$$\rho g y \, dy = \sigma \sin\theta \, d\theta. \tag{5.33}$$

Integrating this equation, with the boundary condition that $\theta = 0$ at $y = 0$, Eq. (5.33) becomes

$$\rho g \int_0^y y\,dy = \sigma \int_0^\theta \sin\theta\,d\theta. \tag{5.34}$$

Evaluating the integrals,

$$\tfrac{1}{2}\rho g y^2 = \sigma(1 - \cos\theta). \tag{5.35}$$

Substituting $2\sin^2 \theta/2 = (1 - \cos\theta)$,

$$y^2 = \left(\frac{4\sigma}{\rho g}\right) \sin^2 \frac{\theta}{2}. \tag{5.36}$$

That is, since we require the positive solution,

$$y = 2\left(\frac{\sigma}{\rho g}\right)^{1/2} \sin \frac{\theta}{2}. \tag{5.37}$$

This gives the surface on the left hand side of the plate (Fig. 5.5). The same derivation can be applied to the surface on the right hand side of the plate. However the angle θ, defined for the right hand surface, is negative as indicated in Fig. 5.5. Substituting $(-\theta)$ for θ in Eq. (5.36) gives

$$y = 2\left(\frac{\sigma}{\rho g}\right)^{1/2} \sin \frac{\theta}{2}. \tag{5.38}$$

The height to which the liquid rises up the plate on the left hand side, y_l, is obtained by substituting $\theta = \beta - \alpha$ into (5.37). This gives

$$y_l = 2\left(\frac{\sigma}{\rho g}\right)^{1/2} \sin \tfrac{1}{2}(\beta - \alpha). \tag{5.39}$$

The height to which the liquid rises on the right hand side, y_r, is obtained by putting $\theta = \pi - \beta - \alpha$ in 5.38, giving

$$y_r = 2\left(\frac{\sigma}{\rho g}\right)^{1/2} \cos \tfrac{1}{2}(\beta + \alpha). \tag{5.40}$$

The liquid rises to different heights on each side of the plate.

The magnitude of the rise in height of the liquid up the plate can be gauged by evaluating the factor $2(\sigma/\rho g)^{1/2}$. For water and soap solution at 20 °C, σ has values of approximately 73 and 32 dynes per cm respectively. Substituting $\rho = 1$ gm per cc this factor is 5.4 mm and 3.6 mm respectively.

The Eq. (5.40) is valid for all values of β. In the case $\beta = 0$ the plate is horizontal and above the level of the liquid at infinity. The liquid surface rises to a height y_0 given by,

$$y_0 = 2\left(\frac{\sigma}{\rho g}\right)^{1/2} \cos\left(\frac{\alpha}{2}\right). \tag{5.41}$$

This is shown in Fig. 5.6.

Fig. 5.6   The limiting case of y_r with $\beta = 0$.

Equation (5.37) can be expressed in Cartesian coordinates (x, y) by substituting

$$\frac{dy}{dx} = \tan \theta. \tag{5.42}$$

This leads to a differential equation for $y(x)$ which can only be solved numerically. So it is more convenient to use coordinates (θ, y) to examine the surface configuration analytically.

5.4   The Rise of Liquid between Parallel Vertical Plates

The rise of liquid between two parallel vertical plates, shown in Fig. 5.7(a), can be analysed in the same way as the rise of fluid up an inclined plate. Equation (5.33) determines the surface configuration. In this case, however, the fluid does not extend to infinity but is bounded by the plates on either side. So the boundary conditions are different from the case of the single inclined plane.

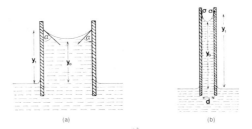

(a)   (b)

Fig. 5.7(a)   The rise of liquid between two parallel vertical plates. (b) Two parallel plates separated by a small distance, d.

Let the surface of the liquid midway between the plates rise to a height y_0. Midway between the plates $\theta = 0$ by symmetry. Integrating Eq. (5.33),

$$\rho g \int_{y_0}^{y} y \, dy = \sigma \int_{0}^{\theta} \sin \theta \, d\theta. \qquad (5.43)$$

Hence

$$\tfrac{1}{2}\rho g(y^2 - y_0^2) = \sigma(1 - \cos \theta). \qquad (5.44)$$

Alternatively,

$$y^2 - y_0^2 = 4\left(\frac{\sigma}{\rho g}\right) \sin^2 \frac{\theta}{2}. \qquad (5.45)$$

If the angle of contact at both plates is α and the height of the surface at the plates is y_1, Eq. (5.45) gives

$$y_1^2 - y_0^2 = 4\left(\frac{\sigma}{\rho g}\right) \sin^2\left(\frac{\pi}{4} - \frac{\alpha}{2}\right), \qquad (5.46)$$

as $\theta = \pi/2 - \alpha$ at the boundary.

Equation (5.45) gives the shape of the surface between the two vertical plates. It does not give the absolute height to which the liquid rises. In order to determine this it is necessary to consider the equilibrium of the column of liquid. The vertical force due to the surface tension, at the top of the plates, must balance the weight of the column of liquid. The exact calculation of the weight of the column of liquid cannot easily be carried out analytically as the volume of liquid must be determined. This requires the evaluation of

$$\int_0^d y\,dx, \tag{5.47}$$

where d is the distance between the plates. To evaluate (5.47) y must be expressed as a function of x using (5.45) and (5.42).

In order to indicate the method for the determination of the absolute rise in height of the liquid column let us consider a special case. This is the case of the rise of liquid between plates that are close together. We shall also simplify the problem by assuming that the liquid wets the plates, so the angle of contact is zero (Fig. 5.7(b)).

The Laplace–Young equation for the surface is, from Eq. (5.29),

$$\rho g y = \frac{\sigma}{R_1}, \tag{5.48}$$

where y is the rise in height of a point in the surface with radius of curvature R_1, in the plane perpendicular to the plates. When the plates are close together y is large compared with both the distance between the plates and the distance between the top and bottom of the meniscus, $(y_1 - y_0)$. Thus the variation in y over the surface of the liquid is small. So we are justified assuming that y is constant in Eq. (5.48). Hence we deduce from (5.48) that R_1 is constant. The cross-section of the surface can thus be assumed to be circular. As the contact angle is assumed to be zero, the shape of the surface will be a semicircle of diameter, d, equal to the distance between the plates. The volume contained between a unit width of the plates is

$$y_1 d - \tfrac{1}{2}\pi \left(\frac{d}{2}\right)^2. \tag{5.49}$$

The weight of this volume of liquid is balanced by the surface tension force acting on a unit width of each plate. Hence

$$\rho g\left(y_1 d - \frac{1}{8}\pi d^2\right) = 2\sigma. \tag{5.50}$$

Thus the absolute height, y_1, is given by

$$y_1 = \frac{2\sigma}{\rho g}\left(\frac{1}{d}\right) + \frac{\pi}{8}d. \tag{5.51}$$

In order to obtain y_0 it is necessary to determine $(y_1 - \frac{1}{2}d)$.

5.5 A Liquid Drop contained between Two Closely Spaced Horizontal Plates

A drop of liquid inserted between two closely spaced plates will draw them together with a considerable force providing the angle of contact is less than $\pi/2$. If the angle of contact is greater than $\pi/2$ the plates will be repelled with an equally large force. This force can be determined by the application of the Laplace–Young equation.

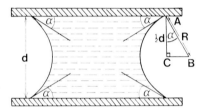

Fig. 5.8   Liquid between two closely spaced horizontal plates.

Figure 5.8 shows a drop of liquid contained between two horizontal plates with an angle of contact less than $\pi/2$. The principal curvature, $(1/R)$, of the surface of the liquid in a plane perpendicular to the plates is shown. The other principal curvature can be neglected as the radius of curvature is much greater than R. The Laplace–Young equation gives the excess pressure, p, across the surface as

$$p = \frac{\sigma}{R}. \tag{5.52}$$

As the pressure inside the fluid is constant, p is constant, Eq. (5.52) gives a constant radius of curvature, R. If the plates are separated by a distance d and the liquid makes an angle of contact α with the plates, R can easily be expressed as a function of d and α. In Fig. 5.8, B is the centre of curvature of the right hand surface. The triangle ABC forms a right angled triangle so that

$$2R\cos\alpha = d. \tag{5.53}$$

So applying (5.52) and (5.53), the pressure inside the liquid is less than the external, atmospheric, pressure by

$$p = \frac{2\sigma \cos a}{d}. \tag{5.54}$$

If S is the area of the drop in contact with each plate and l is the length of the boundary of the liquid in contact with each plate, the total force, f, exerted perpendicular to each plate is composed of a force due to the excess pressure plus a force due to the component of the surface tension force. Thus

$$f = \frac{2\sigma \cos a}{d} S + \sigma l \sin a. \tag{5.55}$$

It follows from Eq. (5.55) that the magnitude of the force between the plates increases with increasing distance, d, between the plates. This force will tend to infinity as d tends to zero. The force will be attractive for a less than $\pi/2$ and repulsive for a greater than $\pi/2$.

A good example of the considerable force that is present for small d occurs when two blocks of ice are placed with their faces close together. If they are allowed to melt slowly, the water formed between them will cause a force sufficient to draw the blocks together. This will cause the blocks to freeze into one block.

The strong repulsive nature of the force between two surfaces is used in the lubrication of moving metallic elements in, for example, a motor car. Using an oil as a lubricant, with $a > \pi/2$, the metallic surfaces are repulsed, thus preventing them from sliding over each other causing frictional wear. The introduction of water in place of oil will have the converse effect. The metallic surfaces will be drawn together with a considerable force and the movement of one element relative to the other will produce a large frictional force resulting in rapid wear and subsequent damage.

When two, almost vertical, plates are inclined to each other at a small angle so that the tops of the plates are closer together than the ends of the plates (Fig. 5.9(a)) liquid contained by the plates will have a radius of curvature r_1 at the top surface which is less than the radius of curvature r_2 at the bottom surface. This is due to the larger pressure difference across the top surface than the bottom surface. The pressure difference across the bottom surface is the pressure difference across the top surface minus that due to the column of liquid of height h. Thus using the Laplace–Young equation (5.52) the relation between r_1 and r_2 at equilibrium is

$$\rho g h = \sigma\left(\frac{1}{r_1} - \frac{1}{r_2}\right). \tag{5.56}$$

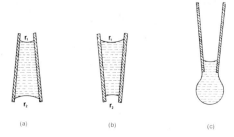

Fig. 5.9   (a) Liquid in equilibrium between two almost vertical, inclined, plates in a Λ-shaped configuration. (b) Non-equilibrium situation for the V-shaped position of the plates. (c) Final equilibrium of liquid between the V-shaped plates.

If the plates are inverted, r_1 at the top surface will be greater than r_2, this equation cannot be satisfied (Fig. 5.9(b)). Equilibrium is not possible in this configuration. The liquid will be drawn to the bottom of the plates and form a pendant drop as shown in Fig. 5.9(c).

5.6   A Large Stationary Drop

A large stationary drop of liquid resting on a horizontal surface will, by symmetry, have a circular cross-section in any horizontal plane. A vertical section, symmetrically through the centre of the drop, is shown in Fig. 5.10. It has an angle of contact α with the horizontal surface on which it rests.

At the centre of the upper surface of the drop, Q, the two principal radii of curvature of the surface will be equal by symmetry. Let these radii be r. The pressure below the surface at Q will exceed that above Q, according to Laplace–Young equation, by

$$\frac{2\sigma}{r}, \tag{5.57}$$

where σ is the surface tension of the drop. If the liquid has a density ρ_2 and that of the surrounding vapour or gas is ρ_1, the pressure difference across a typical point P on the profile of the drop (Fig. 5.10) at a vertical distance y below Q is, from (5.57),

$$\frac{2\sigma}{r} + gy(\rho_2 - \rho_1). \tag{5.58}$$

Applying the Laplace–Young equation at P,

$$\frac{2\sigma}{r} + gy(\rho_2 - \rho_1) = \sigma\left(\frac{1}{R_1} + \frac{1}{R_2}\right), \tag{5.59}$$

where R_1 and R_2 are the principal radii of curvature of the droplet at P. R_1 will be in the plane of Fig. 5.10 and R_2 in a plane perpendicular to this plane. For a large drop $R_2 \gg R_1$. Hence the contribution from $(1/R_2)$ can be neglected in Eq. (5.59). Also, at Q $r \gg R_1$ so the surface can be approximated by a horizontal plane, for which $r = \infty$. Thus the term $(2\sigma/r)$ can also be neglected. Hence (5.59) becomes,

$$\sigma\left(\frac{1}{R_1}\right) = \rho gy, \tag{5.60}$$

where $\rho = \rho_2 - \rho_1$.

This equation is the same as that for a liquid surface produced by an infinite inclined plate which was discussed in section 5.3. It corresponds to

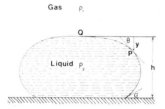

Fig. 5.10   A large stationary drop of liquid resting on a horizontal plane.

the case in which the inclined plate is horizontal and raised above the lower horizontal surface of the drop so that the liquid surface resembles that in Fig. 5.10. Introducing the angle θ between the tangent at P and the horizontal surface through Q, (5.60) becomes

$$\sigma\frac{d\theta}{ds} = \rho gy, \tag{5.61}$$

where ds is an element of the length of the profile of the drop at P. As

$$\frac{dy}{ds} = \sin\theta, \tag{5.62}$$

Eq. (5.61) becomes

$$\sigma \sin \theta \, \frac{d\theta}{dy} = \rho g y. \tag{5.63}$$

At the point Q $\theta = 0$ and $y = 0$, so integrating (5.63),

$$\sigma \int_0^\theta \sin \theta \, d\theta = \rho g \int_0^y y \, dy. \tag{5.64}$$

This gives, on performing the integration,

$$y^2 = \frac{2\sigma}{\rho g}(1 - \cos \theta). \tag{5.65}$$

As $1 - \cos \theta = 2 \sin^2 \theta/2$,

$$y = 2\left(\frac{\sigma}{\rho g}\right)^{1/2} \sin\frac{\theta}{2}. \tag{5.66}$$

If the point Q is at a height h above the horizontal plane on which the drop rests then, as $\theta = \pi - a$ at $y = h$, (5.66) gives

$$h = 2\left(\frac{\sigma}{\rho g}\right)^{1/2} \cos\frac{a}{2}. \tag{5.67}$$

The height, h, depends only on the surface tension, σ, the angle of contact, a, and the density, ρ.

This analysis can also be applied to large gas bubbles in a liquid medium resting on, or under, a horizontal plane or produced by gas from a glass tube inserted into the liquid. Pendant drops attached to the underside of a horizontal plate, or projecting from the lower end of a glass tube, in a gaseous or vapour medium can also be analysed in this way.

When the drops and bubbles are not sufficiently large to justify the approximations that have been made in obtaining Eq. (5.60) the radius of curvature R_2 has to be retained. Problems that need both radii of curvature require the introduction of a further coordinate, in addition to y and θ. This more general case of the Laplace–Young differential equation cannot be solved by analytic methods and one must resort to a numerical solution.

There is another special case of the Laplace–Young equation that is amenable to an analytic solution. This is the case of constant excess pressure across the liquid surface in a capillary tube which results in a spherical surface. This will be examined in the next section.

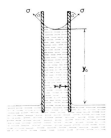

Fig. 5.11   The rise of liquid in a narrow capillary tube.

5.7   The Rise of Liquid in a Narrow Capillary Tube

A narrow capillary tube inserted vertically into a bath of liquid will cause the liquid to rise to a height that is considerably greater than the diameter of the tube (Fig. 5.11). If the liquid has density ρ_2, and that of the surrounding gas or vapour ρ_1, then any point on the surface of the meniscus at a height y above the liquid in the bath, with principal radii of curvature R_1 and R_2, will satisfy the Laplace–Young equation,

$$\rho g y = \sigma \left(\frac{1}{R_1} + \frac{1}{R_2} \right), \tag{5.68}$$

where σ is the surface tension and $\rho = \rho_2 - \rho_1$. The variation of y over the surface of the meniscus is assumed to be small so Eq. (5.68) becomes

$$\frac{1}{R_1} + \frac{1}{R_2} = \text{constant}. \tag{5.69}$$

This is the same equation as that for a spherical droplet. As the boundary of meniscus is circular, the surface is spherical.

The equilibrium of the column of liquid will enable the rise, y_0, of the liquid at the centre of the capillary tube to be determined. The vertical upward force on the column of liquid is due to the surface tension force acting at the circumference of the meniscus. If the contact angle is α, this force is

$$2\pi r \sigma \cos \alpha, \tag{5.70}$$

where r is the radius of the tube. This force is balanced by the downward weight of the column. It consists of the weight of the cylindrical column of height y_0 plus the weight of the meniscus lens of liquid above y_0. This has a volume $\pi r^2 \Delta y_0$ where Δy_0 can be shown to be given by

$$\Delta y_0 = r(\sec a + \tfrac{2}{3} \tan^3 a - \tfrac{2}{3} \sec^3 a). \tag{5.71}$$

Equating the two forces,

$$2\pi r \sigma \cos a = \rho g \pi r^2 (y_0 + \Delta y_0). \tag{5.72}$$

So

$$y_0 = \left(\frac{2\sigma \cos a}{\rho g}\right)\frac{1}{r} - \Delta y_0, \tag{5.73}$$

where Δy_0 is given by (5.71).

If the liquid wets the surface of the tube $a = 0$ and, from (5.71), $\Delta y_0 = \dfrac{r}{3}$. Thus (5.73) becomes,

$$y_0 = \left(\frac{2\sigma}{\rho g}\right)\frac{1}{r} - \frac{r}{3}. \tag{5.74}$$

The rise of liquid in a capillary tube provides a simple method of determining σ, providing the radius of the tube is sufficiently small that the meniscus can be considered as part of a spherical surface. One of the practical problems associated with the measurement of both σ and a is the difficulty in ensuring that both the liquid surface and the glass tube are free from dirt and impurities.

5.8 The Equilibrium of a Pendant Drop

A pendant drop of liquid at the end of a tube is shown in Fig. 5.12(a). PQ is the symmetry axis of the system, P being the centre of the top of the meniscus and Q the centre of the bottom of the drop. The distance PQ is h. At P the principal radii of curvature of the surface will be equal by symmetry. Let their values be R. The liquid at P will have a pressure that is less than that of the surrounding atmosphere by an amount given by the Laplace–Young equation as,

$$\sigma\left(\frac{1}{R} + \frac{1}{R}\right) = \frac{2\sigma}{R}, \tag{5.75}$$

where σ is the surface tension of the liquid.

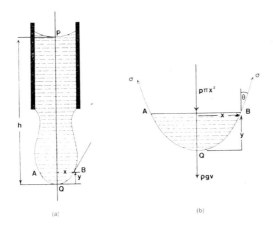

Fig. 5.12   (a) The equilibrium of a pendant drop. (b) A detailed section through the bottom of the pendant drop showing the forces present.

Consider the section of the pendant drop contained within a height y above Q. This is labelled ABQ in Fig. 5.12(b). The liquid is assumed to have a density ρ_2 and the surrounding atmosphere a density ρ_1. The pressure difference between the liquid in the plane AB and the liquid at P is

$$\rho g(h-y), \tag{5.76}$$

where $\rho = \rho_2 - \rho_1$. Thus the pressure difference, p, between the liquid at B and the air at B, from (5.75) and (5.76), is given by

$$p = \rho g(h-y) - \frac{2\sigma}{R}. \tag{5.77}$$

The forces acting on the section of the pendant drop below AB are: (a) the surface tension force acting tangential to the perimeter surface; (b) the force due to the pressure of the column of liquid above AB; (c) the weight of the liquid within ABQ, of volume v. The latter two forces act vertically downwards and (a) has a resultant acting vertically upwards that balances (b) plus (c). If B is at a distance x from the central axis of symmetry and the tangent at B makes an angle θ with the vertical, on equating the vertical forces, using (5.77),

$$2\pi x\sigma\cos\theta = \left(\rho g(h-y) - \frac{2\sigma}{R}\right)\pi x^2 + \rho g v. \tag{5.78}$$

For small y the surface curve, AQB, of the drop can be approximated by a parabola,

$$y = ax^2, \tag{5.79}$$

where a is a constant. The volume of this section of the drop is thus given by,

$$v = \pi \int_0^y x^2 dy, \tag{5.80}$$

from (5.79),

$$= \frac{\pi}{a} \int_0^y y\,dy.$$

Therefore,

$$v = \frac{\pi y^2}{2a}. \tag{5.81}$$

Substituting (5.81) into (5.78),

$$2\pi x\sigma\cos\theta = \left(\rho g(h-y) - \frac{2\sigma}{R}\right)\pi x^2 + \rho g\frac{\pi y^2}{2a}. \tag{5.82}$$

Collecting terms in σ,

$$2\sigma x\left(\cos\theta + \frac{x}{R}\right) = \rho g\left(\frac{y^2}{2a} + x^2(h-y)\right). \tag{5.83}$$

Finally, replacing a by (y/x^2), we obtain an expression for σ,

$$\sigma = \frac{\rho g x h}{2}\left(\frac{1-(y/2h)}{\cos\theta + (x/R)}\right). \tag{5.84}$$

5.9   The Influence of Surface Tension on Gravity Waves

The velocity, c, of 'gravity waves' propagating over the surface of a deep liquid is related to their wavelength, λ, by[30]

$$c^2 = \frac{g\lambda}{2\pi}.$$ (5.85)

This result assumes that the effects due to surface tension can be neglected. The depth of the liquid is also required to be much greater than λ.

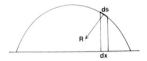

Fig. 5.13   Profile of a section of a wave.

When the influence of surface tension cannot be neglected it is necessary to generalize the result (5.85) for the velocity of propagation of the waves in order to account for surface forces. Waves in which surface tension forces cannot be neglected are called *capillary waves*. Consider a plane wave with amplitude A moving along the x-axis, in Cartesian coordinates, having a profile of the form,

$$y = A \sin\left(\frac{2\pi}{\lambda} x\right),$$ (5.86)

at a fixed instant of time. Figure 5.13 shows part of the wave profile. At any point on this surface the principal radius of curvature will be zero in the plane perpendicular to the figure and non zero, equal to R, in the plane of the figure. The excess pressure inside the surface of the wave is, according to the Laplace–Young equation,

$$\frac{\sigma}{R},$$ (5.87)

where σ is the surface tension of the liquid. In Cartesian coordinates R is given by,

$$R = \frac{\left[1 + \left(\frac{dy}{dx}\right)^2\right]^{3/2}}{\frac{d^2y}{dx^2}}.$$ (5.88)

Under normal conditions we are justified in assuming that $dy/dx \ll 1$, so (5.88) becomes

$$R = \frac{1}{\dfrac{d^2y}{dx^2}}. \tag{5.89}$$

From (5.89) and (5.86),

$$\frac{1}{R} = \frac{d^2y}{dx^2}, \tag{5.90}$$

$$= -\frac{4\pi^2}{\lambda^2}\, y. \tag{5.91}$$

Thus the magnitude of the excess force on a small element of the surface, PQ, of length ds and of unit width is

$$\frac{4\pi^2}{\lambda^2}\, y\sigma\, ds. \tag{5.92}$$

Resolving this force vertically downwards, through an angle of $\cos^{-1}(dx/ds)$,

$$\frac{4\pi^2}{\lambda^2} y\sigma\, ds \left(\frac{dx}{ds}\right) = \frac{4\pi^2}{\lambda^2} y\sigma\, dx. \tag{5.93}$$

The total vertical force due to the weight of liquid above the mean level of the surface plus the force due to the excess pressure is

$$\rho g y\, dx + \left(\frac{4\pi^2}{\lambda^2}\right)\sigma y\, dx, \tag{5.94}$$

$$= y\rho\left(g + \frac{4\pi^2\sigma}{\lambda^2\rho}\right)dx. \tag{5.95}$$

The surface tension thus increases the effect of g in (5.85) to a new effective value of

$$g + \frac{4\pi^2\sigma}{\lambda^2\rho}. \tag{5.96}$$

So the new velocity of propagation of the waves, Eq. (5.85), becomes,

$$c^2 = \frac{\lambda}{2\pi}\left(g + \frac{4\pi^2\sigma}{\lambda^2\rho}\right).$$

(5.97)

When

$$g \gg \frac{4\pi^2\sigma}{\lambda^2\rho},$$

(5.98)

the effect of the surface tension force can be neglected. However for sufficiently small wavelengths,

$$\frac{4\pi^2\sigma}{\lambda^2\rho} \gg g,$$

(5.99)

the surface tension force is dominant and (5.97) gives

$$c^2 = \frac{2\pi\sigma}{\lambda\rho}.$$

(5.100)

From (5.97) c has a minimum value, for a variation in λ, when

$$\lambda^2 = \frac{4\pi^2\sigma}{\rho g}.$$

(5.101)

For water waves this occurs at $\lambda = 1.7$ cm and $c = 23$ cm per sec. It is conventional to call wavelengths that are smaller than the minimum value, *ripples*, and longer wavelengths simply *waves*.

5.10   Liquid Jets and Cylindrical Films

A jet of water issuing from a circular orifice, or tap, does not retain its cylindrical form. The jet of water will break up into droplets. A careful examination of the jet will show that oscillations appear in the surface of the jet which grow in amplitude along the trajectory of the jet, and finally cause the jet to break up into droplets (Fig. 5.14(a)). Similar phenomena are to be observed in closed stationary cylinders of soap film. The surface develops oscillations and eventually breaks up into soap bubbles if the cylinder is sufficiently long (Fig. 5.14(b)). However if the length of the cylinder of soap film is small enough any oscillations that occur in the surface will decay away.

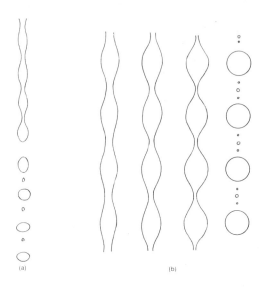

Fig. 5.14 (a) An unstable liquid jet. (b) The growth of an instability in a cylindrical soap film leading to the break-up of the cylinder into bubbles.

These instabilities can be explained by considering the pressure variations in the jet or cylindrical soap film. Consider a cylinder of fluid, or cylindrical soap film, of radius a (Fig. 5.15). Let the axis of the cylinder be the x-axis and the radial axis be the y-axis. Now consider the stability of the cylinder when it is perturbed by a small periodic disturbance of wavelength, λ, so that the cylinder is deformed from $y = a$ to,

$$y = a + b \cos\left(\frac{2\pi x}{\lambda}\right), \qquad (5.102)$$

where $b \ll a$. The system will be stable if the pressure at the region of greatest radial extension, such as A, is greater than that at the region of greatest radial contraction, such as B. The liquid in the case of the jet, or air in the case of the cylindrical film, will then flow from the high pressure region to the low pressure region in order to restore the cylindrical shape. To determine the conditions under which the cylinder is stable it is necessary to calculate this pressure difference using the Laplace–Young equation.

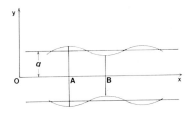

Fig. 5.15 A perturbed cylinder of liquid.

At any point on the surface such as A the principal radii of curvature, R_1 and R_2, determine the excess pressure in the liquid. R_1 will be the maximum radius of the 'cylindrical' surface. From (5.102),

$$R_1 = a+b. \tag{5.103}$$

R_2 is given by

$$R_2 = \frac{\left[1 + \left(\dfrac{dy}{dx}\right)^2\right]^{3/2}}{\dfrac{d^2y}{dx^2}}. \tag{5.104}$$

For small deformations $dy/dx \ll 1$ so (5.104) becomes

$$R_2 = \frac{1}{\dfrac{d^2y}{dx^2}}. \tag{5.105}$$

Differentiating (5.102), the magnitude of R_2 at the point of greatest extension is given by

$$\frac{1}{R_2} = \frac{4\pi^2b}{\lambda^2}. \tag{5.106}$$

Thus from (5.103) and (5.106) the pressure of the liquid at A is greater than the external pressure by

$$\sigma\left(\frac{1}{R_1} + \frac{1}{R_2}\right) = \sigma\left(\frac{1}{a+b} + \frac{4\pi^2 b}{\lambda^2}\right). \tag{5.107}$$

Similarly the pressure difference of the liquid at B is greater than the external pressure by

$$\sigma\left(\frac{1}{a-b} - \frac{4\pi^2 b}{\lambda^2}\right). \tag{5.108}$$

In (5.107) and (5.108) we have assumed that a liquid cylinder has been deformed. In the case of the cylinder of soap film Eq. (5.108) is valid providing σ is the film tension, σ_f. This factor will not effect the condition we shall derive for the stability of the soap film. Neglecting the weight of the fluid, the pressure difference between A and B is, from (5.107) and (5.108),

$$\sigma\left(\frac{1}{a+b} - \frac{1}{a-b} + \frac{8\pi^2 b}{\lambda^2}\right), \tag{5.109}$$

$$= \sigma\left(\frac{8\pi^2 b}{\lambda^2} - \frac{2b}{a^2 - b^2}\right). \tag{5.110}$$

As $b \ll a$ this reduces to

$$\sigma\left(\frac{8\pi^2 b}{\lambda^2} - \frac{2b}{a^2}\right), \tag{5.111}$$

$$= \frac{2b\sigma}{a^2}\left[\left(\frac{2\pi a}{\lambda}\right)^2 - 1\right]. \tag{5.112}$$

For the liquid jet, or soap film, to be stable this pressure difference must be positive. That is

$$2\pi a > \lambda. \tag{5.113}$$

If this condition is satisfied any disturbance, or perturbation, will decay away and the jet, or film, will return to its cylindrical shape. The cylinder becomes unstable if $\lambda \geqslant 2\pi a$. Thus a small perturbation will grow if $\lambda \geqslant 2\pi a$ and eventually cause the jet, or soap film, to break up. This condition was originally derived by Lord Rayleigh.[54] He also showed that the maximum instability occurs when $\lambda = 9.02a$.

The behaviour of an unstable jet of fluid, $\lambda \geqslant 2\pi a$, is shown in Fig. 5.14(a). A chance disturbance imposed on the jet causes its profile to undergo oscillations which build up with time and eventually lead to the break up of the jet into separate droplets which tend to oscillate due to their initial distortion. In the case of a cylindrical soap film (Fig. 5.14(b)) a similar sequence occurs. The oscillations grow with time and eventually cause the soap film to break up into individual bubbles.

Water issuing from a tap is probably the commonest example of this instability. However there are other examples which can be observed. For example a spider when building its web produces a cylindrical tube of sticky liquid that wets the web. This liquid cylinder, due to its inherent instability, will break up into beads of sticky liquid. These beads are about 10^{-3} cm in radius and are usually of two kinds arranged alternately along the web. One type of bead is about three times the diameter of the other and arranged periodically along the web. The two types of beads are due to the two characteristic wavelengths that evolve along the jet profile. The sticky matter is used to trap flies.

Another example occurs in the action of fuse wire. When the electrical heating is sufficiently great it will cause the wire to melt and form a liquid cylinder. The cylinder will be unstable, break up into droplets of liquid, and thus break the electrical circuit. The effect of gravitational force on the liquid fuse wire will also tend to break the electrical connection, however it can be neglected as with the speed of propagation the surface instability is much greater than the speed of the gravitational instability.

These instabilities have usually been initiated by random fluctuations with the property that $\lambda \geqslant 2\pi a$. It is possible to impose a perturbation on a cylindrical jet, or cylinder of soap film, by using a frequency source such as a tuning fork. If the tuning fork provides perturbing oscillations that satisfy the condition $\lambda \geqslant 2\pi a$, it will cause the jet, or film, to become unstable and break up. Two perturbing tuning forks of different frequencies, that satisfy the instability criterion, will produce two characteristic oscillations of the cylinder. This will cause the fluid cylinder to break up into two sets of drops. Each set of drops being associated with one of the perturbing frequencies.

These sets of droplets from a jet will have different velocities and masses, forming two different trajectories. This is most noticeable if the jet is inclined to the vertical so that the two trajectories form two different parabolic paths.

Further perturbing frequencies will produce more trajectories. Drops from different trajectories that do collide will tend to bounce off each other.

The stability of the jet and the behaviour of the droplets are sensitive to electrical fields and to changes in surface tension. The presence of a static electric field, or the addition of a chemical to the jet liquid, can cause the drops forming a trajectory to coalesce again.

6 ANALYTIC METHODS AND RESULTS, VIBRATIONAL MODES, AND FURTHER ANALOGUE METHODS

6.1 The Calculus of Variations

At this stage it is worthwhile describing, qualitatively, the analytic methods developed by Euler and Lagrange and discussing some of the results. The detailed analysis associated with the solution of some problems requiring the determination of the minimum area contained by fixed boundaries is given in the appendices.

The methods developed by Euler and Lagrange in the eighteenth century come under the general title of *The Calculus of Variations*. These methods are best illustrated by considering the simplest minimization problem, the problem of determining the minimum length of path joining two points.

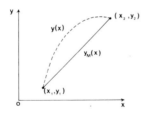

Fig. 6.1   The minimum path joining two points, $y_M(x)$, and a varied path, $y(x)$.

Let the two points be (x_1, y_1) and (x_2, y_2) in Cartesian coordinates. It is assumed that there exists a minimum path, $y_M(x)$, that passes through (x_1, y_1) and (x_2, y_2). Then their method of solution examines a sequence of paths that deviate slightly from the minimum path and pass through the two end points. This is illustrated in Fig. 6.1. This sequence of paths can be written as

$$y(x) = y_M(x) + a\eta(x), \tag{6.1}$$

137

where a is a small parameter that enables us to vary the path, and $\eta(x)$ is an arbitrary function of x that is zero at x_1 and x_2. The varied path, $y(x)$, thus passes through (x_1, y_1) and (x_2, y_2) and gives the minimum path, y_M, when $a = 0$.

It is required to minimize the length of path joining (x_1, y_1) and (x_2, y_2). If ds is an element of the path, the total length of path, L, for any varied path, (6.1), is given by

$$L = \int_{(x_1, y_1)}^{(x_2, y_2)} ds. \tag{6.2}$$

In Cartesian coordinates, using Pythagoras's theorem,

$$ds^2 = dx^2 + dy^2. \tag{6.3}$$

Substituting into (6.2),

$$L = \int_{x_1}^{x_2} \left[1 + \left(\frac{dy}{dx}\right)^2\right]^{1/2} dx, \tag{6.4}$$

$$= \int_{x_1}^{x_2} (1 + y_x^2)^{1/2} dx, \tag{6.5}$$

where

$$y_x = \frac{dy}{dx}. \tag{6.6}$$

The length, L, depends on the integrand in (6.5), and hence, from (6.6) and (6.1), on the parameter, a, that specifies the varied path from (x_1, y_1) to (x_2, y_2).

Using the methods of the calculus, it is shown in Appendix I that the condition for the more general integral

$$J = \int_{x_1}^{x_2} f(x, y, y_x) dx, \tag{6.7}$$

where $f(x, y, y_x)$ is a function of x, y, and y_x, to be an extremum is that f must satisfy the Euler–Lagrange differential equation,

$$\frac{d}{dx}\left(\frac{\partial f}{\partial y_x}\right) - \frac{\partial f}{\partial y} = 0. \tag{6.8}$$

The function $f(x, y, y_x)$ is known. Consequently this equation enables the path, for which J is an extremum, to be determined. In the example of the path joining (x_1, y_1) and (x_2, y_2) f is given by the integrand in (6.5). The solution to the differential Eq. (6.8) gives the extremum path, y_M.

The derivation of Eq. (6.8) assumes that the extremum curve is continuous and has a continuous derivative. Consequently it cannot be used to solve problems which have a discontinuous solution or a discontinuous derivative. Such problems as the Steiner problems and the minimum area problems, discussed in Chapters 3 and 4, can have discontinuous gradients and consequently cannot be solved using the Euler–Lagrange equation.

The Euler–Lagrange equation, (6.8), is derived in Appendix I and applied, in Appendix II, to the problem of the determination of the minimum length of path joining two points. Appendix III contains the detailed solution to the problem of the surface of minimum area contained by two coaxial parallel rings situated perpendicular to their axis. The solution surface is a catenoid, providing the rings are sufficiently close together, or two discs. The detailed analysis shows that there is always another solution formed by two discs contained by the rings, with no intermediate surface. If the rings are sufficiently close together the two discs form a surface which has an area that is a relative minimum, the catenoid being the surface with the absolute minimum area. However as the rings are drawn apart a stage is reached at which the area of the catenoid becomes greater than that of the discs. Consequently the two discs form the surface with the absolute minimum area and the catenoid area is only a relative minimum. Finally on separating the rings further a critical separation is reached at which the minimum area catenoid ceases to exist. At this distance, and for further separation of the rings, the two discs form the only minimum surface.

In Appendix IV the Euler–Lagrange equation is derived for the general two dimensional problem. It is applied to the minimum area contained by a simple closed line boundary. It is shown analytically that the surface area will be minimized if every point on the surface, with principal radii of curvature R_1 and R_2, satisfies the equation,

$$\frac{1}{R_1} + \frac{1}{R_2} = 0. \tag{6.9}$$

This is the Laplace–Young equation for a surface with zero pressure difference. The method can be extended to the general problem of the minimization of the free energy and leads to result

$$\frac{1}{R_1} + \frac{1}{R_2} = \frac{p_f}{\sigma_f}, \tag{6.10}$$

for any point on the surface with an excess pressure, p_f, and film tension, σ_f.

The calculus of variations is applied, in Appendix V, to the problem of the maximum area occupied by a closed perimeter of fixed length. The perimeter forms the circumference of a circle in order to maximize the area. This was demonstrated experimentally, in section 1.3, using a loose loop of thread placed inside a ring containing a disc of soap film. The soap film inside the loop was broken with the result that the loop was pulled by the surrounding soap film into a circle (Plates 1.1(a) and (b)). The hole formed inside the loop takes up its maximum area because the area of the surrounding soap film is minimized. It can also be shown that if the area remains fixed and the perimeter is allowed to vary the minimum length of perimeter occurs when the area is contained by a circle.

The three dimensional generalization of this result is obtained in Appendix VI. That is, the maximum volume contained by a simple closed surface of fixed area is obtained when the volume forms a sphere. It can also be proved that, for a fixed volume, the surface area is minimized when the volume occupies a sphere. This result can be demonstrated with a soap bubble. If the volume of gas in the bubble remains constant and the surface is distorted, the minimum total free energy of the system results when the system reaches equilibrium. This occurs when the surface area is minimized and the bubble has the shape of a sphere.

Some important mathematical results have recently been proved by Frederick J. Almgren Jr. and Jean E. Taylor. They have been able to show that Plateau's rules for the angles of intersection of soap film surfaces and lines, discussed in Chapter 4, are a general property of the solution of the Laplace–Young equation. These proofs are beyond the scope of this book but references to the work can be found in the popular article by Almgren and Taylor[78] entitled *The Geometry of Soap Films and Soap Bubbles.*

6.2 Vibrations of a Soap Film Membrane

A soap film has many properties in common with an elastic membrane. If it is contained by a circular ring it can be made to vibrate transversely, in a direction perpendicular to the ring, like a drum skin. These vibrations are similar to those produced by an elastic string which is fixed at both ends. The string has a number of characteristic vibrations in which all elements of the string vibrate perpendicular to the string with a fixed frequency. These are known as the normal mode frequencies (Plates 6.1(a) and (b)). The special motions associated with all the normal mode frequencies can be used to describe the motion of the string under any circumstances.

A membrane is a 'two dimensional string'. There are special frequencies of vibration, normal modes, of the membrane in which all elements vibrate with the same angular frequency, ω. Once all of these special motions are known it is possible to describe any general motion of the membrane.[31] We shall begin the study of the vibrations of membranes by obtaining the equation of motion of the soap film membrane and hence obtain the normal mode frequencies.

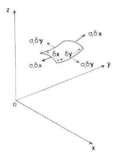

Fig. 6.2   An element of a soap film   Fig. 6.3   A $(y–z)$ section of the element
            membrane.   of the soap film.

Consider a soap film membrane, with film tension σ_f, contained by a closed wire boundary in the $(x–y)$ plane which can vibrate transversely in the z-direction (Fig. 6.2). A small rectangular element of the film at (x, y), of length δx and width δy, will experience forces $\sigma_f \delta x$ acting tangential to the surface in a $y–z$ plane maintaining the element in tension (Fig. 6.2). There will also be a similar force, $\sigma_f \delta y$, acting tangentially to the surface in a $x–z$ plane. The element has constant surface density, ρ, and will move in the z-direction with a force equal to the product of its mass and its acceleration. That is

$$\rho \, \delta x \, \delta y \, \frac{\partial^2 z}{\partial t^2}, \qquad (6.11)$$

where t is the time variable. This force will be produced by the z-component of the surface tension forces. Figure 6.3 is a section of the film in a $(y–z)$ plane. The forces $\sigma_f \delta x$ make angles θ, at x, and $(\theta + \delta\theta)$, at $x + \delta x$, with the $x–y$ plane. Thus the resultant force in the z-direction is

$$\sigma_f \delta x \sin(\theta + \delta\theta) - \sigma_f \delta x \sin \theta. \tag{6.12}$$

As θ is assumed to be small,

$$\tan \theta \simeq \sin \theta. \tag{6.13}$$

So (6.12) becomes,

$$\sigma_f \delta x (\tan(\theta + \delta\theta) - \tan \theta). \tag{6.14}$$

Now $\tan \theta$ is the gradient $\partial z / \partial y$, so (6.14) becomes,

$$\sigma_f \delta x \left[\left(\frac{\partial z}{\partial y} \right)_{y+\delta y} - \left(\frac{\partial z}{\partial y} \right)_y \right]. \tag{6.15}$$

On expanding the first term in brackets we have, to lowest order in δy,

$$\sigma_f \delta x \frac{\partial}{\partial y} \left(\frac{\partial z}{\partial y} \right) \delta y, \tag{6.16}$$

$$= \sigma_f \delta x \delta y \frac{\partial^2 z}{\partial y^2}. \tag{6.17}$$

Similarly for the forces $\sigma_f \delta y$ in the $x–z$ plane, the resultant force in the z-direction is

$$\sigma_f \delta x \delta y \frac{\partial^2 z}{\partial x^2}. \tag{6.18}$$

The total force in the z-direction, from (6.17) and (6.18), is

$$\sigma_f \delta x \delta y \left(\frac{\partial^2 z}{\partial x^2} + \frac{\partial^2 z}{\partial y^2} \right). \tag{6.19}$$

Equating (6.11) and (6.19),

$$\frac{\partial^2 z}{\partial x^2} + \frac{\partial^2 z}{\partial y^2} = \left(\frac{\rho}{\sigma_f}\right) \frac{\partial^2 z}{\partial t^2}, \tag{6.20}$$

or

$$\frac{\partial^2 z}{\partial x^2} + \frac{\partial^2 z}{\partial y^2} = \frac{1}{c^2} \frac{\partial^2 z}{\partial t^2}, \tag{6.21}$$

where

$$c^2 = \frac{\sigma_f}{\rho}. \tag{6.22}$$

Equation (6.21) is the two dimensional wave equation. For an elastic string the term $\partial^2 z/\partial y^2$ is absent and (6.21) reduces to the one dimensional wave equation.

6.3   Normal Modes of a Rectangular Membrane

The normal modes of vibration of a membrane are the solutions of the differential Eq. (6.21) for which

$$z = W(x,y)\exp(i\omega t), \tag{6.23}$$

where $W(x,y)$ is a function of x and y. In these motions the point at (x,y) vibrates in the z-direction with constant frequency, ω. The normal mode angular frequencies, ω, and the vibrational pattern $W(x,y)$ will depend on the boundary conditions imposed on the soap film. We shall examine two examples: a rectangular boundary; and a circular boundary.

First let us consider a rectangular boundary of length, a, and width, b, in the (x,y) plane bounded by the lines

$$x = 0, y = 0, x = a, \text{ and } y = b, \tag{6.24}$$

where a and b are constants.

The solutions, of the form (6.23), that satisfy (6.21) are

$$z = z_0 \frac{\sin}{\cos} px \frac{\sin}{\cos} qy \exp(i\omega t), \tag{6.25}$$

where

$$p^2 + q^2 = \frac{\omega^2}{c^2}, \tag{6.26}$$

and z_0 is a constant. The notation is that in which

$$\frac{\sin}{\cos} \theta = A \sin \theta + B \cos \theta, \tag{6.27}$$

where A and B are arbitrary constants.

As the boundary conditions require that

$$z = 0, \text{ at } x = 0 \text{ and } y = 0, \tag{6.28}$$

the cosine terms in (6.25) must be absent, so

$$z = z_0 \sin px \sin qy \exp(i\omega t). \tag{6.29}$$

Further $z = 0$ at $x = a$, so (6.29) requires

$$\sin pa = 0,$$

or

$$p = \frac{m\pi}{a}, \tag{6.30}$$

where m is a positive integer.

Similarly $z = 0$ at $y = b$, so (6.29) gives

$$\sin qb = 0, \tag{6.31}$$

or

$$q = \frac{n\pi}{b}, \tag{6.32}$$

where n is a positive integer.

Thus, from (6.22), (6.26), (6.30) and (6.32), the normal mode frequencies of the rectangualr soap film membrane are given by,

$$\omega^2 = \left(\frac{m^2}{a^2} + \frac{n^2}{b^2}\right) \frac{\pi^2 \sigma_f}{\rho}, \tag{6.33}$$

where $m = 1, 2, 3 \ldots$ and $n = 1, 2, 3 \ldots$

The (m, n)th mode will have an amplitude at the point (x, y), from (6.29), of

$$z_0 \sin px \sin qy. \tag{6.34}$$

Thus the lines in the $(x-y)$ plane with

$$x = 0, \frac{a}{m}, \frac{2a}{m}, \ldots a, \qquad (6.35)$$

and

$$y = 0, \frac{b}{n}, \frac{2b}{n}, \ldots b, \qquad (6.36)$$

will have zero displacement. These are called the nodal lines. On opposite sides of a nodal line the displacements of elements of the membrane have opposite sign. Figure 6.4 shows the amplitude picture associated with some of the normal modes. The shaded areas have displacements which are opposite in sign to the clear areas. The nodal line pictures are known as Chladni figures.

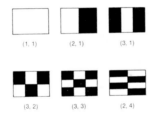

Fig. 6.4   Normal mode patterns of a rectangular membrane labelled by (m, n). The shaded regions are $180°$ out of phase with the clear regions. The horizontal and vertical lines are the nodal lines.

6.4   Normal Modes of a Circular Membrane

A circular membrane, bounded by a ring of radius a, also satisfies the wave Eq. (6.21). It is convenient to use a system of coordinates with the symmetry of the membrane. That is cylindrical coordinates (r, θ, z). The normal mode solutions to (6.21), in cylindrical coordinates, are of the form

$$z = \Lambda(r, \theta) \exp(i\omega t), \qquad (6.37)$$

where $\Lambda(r, \theta)$ is a function of r and θ. The solutions to the wave equation in cylindrical coordinates are

$$z = z_0 J_m(nr) \frac{\sin}{\cos} m\theta \exp(i\omega t), \qquad (6.38)$$

where n is given by $\omega = nc$, m is a constant, J_m is the Bessel function of order m, and z_0 is a constant. The θ-axis can always be chosen such that (6.38) has the form

$$z = z_0 J_m(nr) \cos m\theta \, \exp(i\omega t). \qquad (6.39)$$

As z must be invariant to the addition of a multiple of 2π to θ, m must be an integer that is greater or equal to zero.

At the boundary, $r = a$, $z = 0$. So (6.39) becomes,

$$J_m(na) \cos m\theta \, \exp(i\omega t) = 0, \qquad (6.40)$$

or

$$J_m(na) = 0. \qquad (6.41)$$

For each value of m in (6.41) there are an infinity of values of n that satisfy (6.41). Let us number the roots of this equation with $k = 1, 2, 3 \ldots$ in order of increasing values of n.

The normal mode frequencies are determined by the pair of numbers (m, k). If, for a particular m and k, n has the value $n_{m,k}$, the normal mode frequency is

$$\omega = n_{m,k} c. \qquad (6.42)$$

The nodal lines of the normal mode profile are given by

$$J_m(nr) \cos m\theta = 0. \qquad (6.43)$$

That is

$$J_m(nr) = 0, \qquad (6.44)$$

and for $m \neq 0$,

$$\cos m\theta = 0. \qquad (6.45)$$

That is,

$$\theta = \frac{\pi}{2m} + \frac{s\pi}{m}, \qquad (6.46)$$

where $s = 0, 1, 2, \ldots (2m - 1)$.

Equation (6.44) has solutions for certain fixed values of r. So the nodal lines described by this equation are a series of circles. Equation (6.45), for $m \neq 0$, has solutions for a series of equally spaced values of θ. The nodal lines resulting from (6.45) form a set of radial lines each separated from its neighbour by an angle of π/m. For $m = 0$ the only equation for the nodal lines is (6.44). Some of the normal mode configurations, for the lower frequencies, are given in Fig. 6.5.

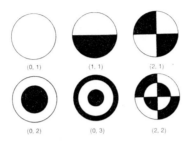

(0, 1)   (1, 1)   (2, 1)

(0, 2)   (0, 3)   (2, 2)

Fig. 6.5   Normal mode patterns of a circular membrane labelled by (m, k). The shaded regions are 180° out of phase with the clear regions. The radial lines and circles are the nodal lines.

The determination of the normal modes of a membrane can be applied to membranes contained by any boundary. It can be extended to problems such as that of a 'kettledrum', in which air is trapped in the drum by a soap film membrane. The motion of the air in the drum must also be taken into account in this problem.

6.5   Normal Mode Experiments

The normal modes of a soap film membrane with a rectangular boundary, circular boundary, or a boundary of any shape, can be easily demonstrated experimentally.[29, 96, 114] It is an advantage to use one of the long lasting recipes, section 1.7, for the soap film. However, solutions made from washing-up liquid can be used but will not have as long a lifetime.

The lower frequency modes can often be demonstrated by hand. A wire frame of any shape can be oscillated or vibrated at the appropriate normal mode frequency to produce a vibrating film with the characteristic nodal line pattern, such as shown in Figs. 6.4 and 6.5. Frames with the dimensions of order 30 cm are suitable for this purpose. A large low dish, or photographic developing tray, provides a simple container for the soap solution. It is not necessary to have large quantities of soap solution as one has only to ensure that the wire frame is immersed in the flat tray of soap solution in order to form a film bounded by the frame. Plates 6.1(a) and (b) show the two lowest frequency modes of the circular membrane. The higher frequency modes cannot easily be obtained by hand. It is necessary to use an electrically driven vibrator, driven at the appropriate frequency, to produce a normal mode vibrational pattern.

A loudspeaker system, with a variable frequency, has been used to produce the normal modes by Dr. L. Bergmann.[80] He photographed the normal mode patterns by reflecting light from the surface of a soap film with dimensions of a few centimetres. The antinodes show up as bright lines. In this way a large number of normal mode frequencies and patterns can be obtained.

6.6   Analogue Solutions to the Differential Equations of Laplace and Poisson

A uniformly stretched rubber membrane is similar in many respects to a soap film, or the interface between two fluids. It has a uniform tension and the thickness of the membrane is small compared with the dimensions of the surface area. The analysis of section 5.2, which derives the Laplace–Young equation for a fluid interface or soap film, applies equally to a uniformly stretched membrane with a transverse pressure load that is perpendicular to the surface. So the Laplace–Young equation for the membrane is

$$p_m = \sigma_m\left(\frac{1}{R_1} + \frac{1}{R_2}\right), \tag{6.47}$$

where p_m is the pressure difference across the membrane, σ_m is the force per unit length in the membrane, or line tension, and R_1 and R_2 are the principal radii of curvature at any point in the membrane surface.

In section 5.1 it was emphasised that the Laplace–Young equation was a differential equation. Thus once the boundary curve of the membrane, or soap film, has been chosen the solution is unique.

If it is not convenient to obtain the configuration of a membrane with a pressure load distribution, p_m, and line tension, σ_m, one can build an analogue system using a soap film. The Laplace–Young equation for the soap film, using σ_f for the film tension and p_f for the pressure difference across the film, is

$$p_f = \sigma_f\left(\frac{1}{R_1} + \frac{1}{R_2}\right). \tag{6.48}$$

Thus using the same boundary conditions for the soap film and the membrane, one can ensure that the solution to (6.48) is identical to (6.47) by setting

$$\frac{p_f}{\sigma_f} = \frac{p_m}{\sigma_m},$$
(6.49)

for all points on the surface of the soap film.

This is a simple application of the soap film to solve a problem satisfying the same differential equation. An important application of soap films to the analogue solution of problems occurs when the gradient at any point on the soap film is small. The Laplace–Young equation in Cartesian coordinates, Eq. (5.9), for the soap film surface

$$z = f(x, y),$$
(6.50)

is

$$\frac{\dfrac{\partial^2 z}{\partial x^2}\left[1 + \left(\dfrac{\partial z}{\partial y}\right)^2\right] - 2\dfrac{\partial z}{\partial x}\dfrac{\partial z}{\partial y}\dfrac{\partial^2 z}{\partial x \partial y} + \dfrac{\partial^2 z}{\partial y^2}\left[1 + \left(\dfrac{\partial z}{\partial x}\right)^2\right]}{\left[1 + \left(\dfrac{\partial z}{\partial x}\right)^2 + \left(\dfrac{\partial z}{\partial y}\right)^2\right]^{3/2}} = \frac{-p_f(x, y)}{\sigma_f}$$
(6.51)

where the sign convention has now been chosen so that the transverse pressure, $p_f(x, y)$, is in the negative z-direction. As the gradients of the film, or membrane, are small,

$$\frac{\partial z}{\partial x} \ll 1, \quad \text{and} \quad \frac{\partial z}{\partial y} \ll 1.$$
(6.52)

Equation (6.51) thus reduces to

$$\frac{\partial^2 z}{\partial x^2} + \frac{\partial^2 z}{\partial y^2} = -\frac{p_f(x, y)}{\sigma_f}.$$
(6.53)

This is Poisson's equation in two dimensions. When $p_f(x, y) = 0$ Eq. (6.53) reduces to Laplace's equation in two dimensions,

$$\frac{\partial^2 z}{\partial x^2} + \frac{\partial^2 z}{\partial y^2} = 0.$$
(6.54)

The differential equations of Poisson and Laplace occur in many branches of physics, other than the fields of membranes and soap films. We can thus use the soap film as an analogue method of solving these equations. It must be remembered that soap films satisfy (6.53) and (6.54) providing the small gradient conditions, (6.52), are satisfied at all points on the surface.

Let us now recall some of the fields of application of Poisson's and Laplace's equation. In electrostatics, for example, the potential, V, over a surface due to a two dimensional charge density, $\rho(x,y)$, satisfies the equation

$$\frac{\partial^2 V}{\partial x^2} + \frac{\partial^2 V}{\partial y^2} = -\frac{\rho(x,y)}{\epsilon}, \tag{6.55}$$

where ϵ is the dielectric constant. If the charge density is zero, (6.55) reduces to Laplace's equation,

$$\frac{\partial^2 V}{\partial x^2} + \frac{\partial^2 V}{\partial y^2} = 0. \tag{6.56}$$

In the elasticity theory of uniform rods, with a uniform cross-section in the $x-y$ planes, the shear stress, σ_{zx} and σ_{zy}, in the x and y directions, due to torques at the end of the bar, are related to the potential function, ϕ, by

$$\sigma_{zx} = \frac{\partial \phi}{\partial y} \quad \text{and} \quad \sigma_{zy} = -\frac{\partial \phi}{\partial x}, \tag{6.57}$$

where the axis of the bar is along the z-direction. The potential function ϕ, satisfies Poisson's equation,

$$\frac{\partial^2 \phi}{\partial x^2} + \frac{\partial^2 \phi}{\partial y^2} = -2G\theta, \tag{6.58}$$

where G is the modulus of rigidity of the bar and θ is the angle of twist per unit length of the bar.

The equation for conduction of electricity in a metal plate of surface resistivity R is the Poisson's equation,

$$\frac{\partial^2 V}{\partial x^2} + \frac{\partial^2 V}{\partial y^2} = -Ri, \tag{6.59}$$

where $i(x,y)$ is the current and $V(x,y)$ the potential at any point (x,y) in the plate.

The steady state equation for heat conduction in a two dimensional plate also satisfies Laplace's equation. If $\theta(x,y)$ is the temperature distribution over the plate then

$$\frac{\partial^2 \theta}{\partial x^2} + \frac{\partial^2 \theta}{\partial y^2} = 0. \tag{6.60}$$

6.7 The Electrostatic Example

Let us examine the procedure for solving one of the previous differential equations using the soap film analogue. Once the method has been outlined for one particular problem the application to the other problems, with the same form of differential equation, will be clear. The electrostatic potential equation, (6.55), is probably the equation that is most widely known. In order to use the soap film analogue we need to replace V, in (6.55), by 'z' in soap film Eq. (6.53). Let us write,

$$V = \lambda z' \tag{6.61}$$

where λ is a dimensional constant. Then (6.55) becomes,

$$\frac{\partial^2 z'}{\partial x^2} + \frac{\partial^2 z'}{\partial y^2} = -\frac{\rho(x,y)}{\lambda \epsilon}. \tag{6.62}$$

It may also be necessary to change the scale of x and y, of the electrostatic problem, for the convenience of the soap film experiment. Let us make a scale change so that the soap film experiment is performed with coordinates, x' and y' respectively, given by

$$x = \alpha x', \tag{6.63}$$

and

$$y = \alpha y', \tag{6.64}$$

where α is a constant. The equation (6.62) in the new coordinates (x',y') is,

$$\frac{\partial^2 z'}{\partial x'^2} + \frac{\partial^2 z'}{\partial y'^2} = -\frac{\alpha^2 \rho(\alpha x', \alpha y')}{\lambda \epsilon}. \tag{6.65}$$

This is identical to the soap film Eq. (6.53) with coordinates (x', y', z') and with an effective pressure given, from (6.53) and (6.65) by

$$p_f(x', y') = \frac{\sigma_f a^2 \rho(ax', ay')}{\lambda \epsilon}. \tag{6.66}$$

So by subjecting the soap film to a pressure distribution $p_f(x', y')$ we have a soap film with height z' that is related by (6.61) to the electrostatic potential, V, and with coordinates that are simply related to those of the potential problem by (6.63) and (6.64). We must also ensure when setting up the soap film experiment that the boundary conditions satisfied by (6.55) are transferred by (6.61), (6.63) and (6.64) into boundary conditions that are satisfied by the soap film experiment.

If $p_f(x', y')$ is constant, or zero, over the soap film surface it is a simple procedure to set up experimentally the pressure difference using gas, or air, on one side of the film that is at a different pressure than that on the other side. However, if a pressure distribution is required it will be necessary to produce a pressure variation over the surface by means of pressure jets which can be altered to give the required pressure distribution. This is more difficult to set up experimentally.

The gradients of the soap film, $\partial z'/\partial x'$ and $\partial z'/\partial y'$, are related by (6.61), (6.63) and (6.64) to the gradients in the potential, $\partial V/\partial x$ and $\partial V/\partial y$, which are the negative components of the electric field $\mathbf{E}$, $(-E_x)$ and $(-E_y)$, in the x and y directions respectively.

Poisson's equation and Laplace's equation can be similarly solved by this analogue method for any of the problems discussed in section 6.6.

Before the advent of digital computers this analogue method was used to solve Poisson's equation and Laplace's equation. Walter J. Karplus and Walter W. Soroka[39] have discussed these methods in some detail in their book *Analog Methods in Computation and Simulation*. At the time of the First World War it was used in the aircraft industry[91, 117] to solve Eq. (6.58).

These equations can now be solved more rapidly, and to greater accuracy, using digital computers. The analogue methods are only capable of giving an accuracy of a few percent. The soap film solutions do, however, have the advantage of providing a visual picture of the complete solution. The digital solutions usually provide only a numerical output.

6.8   The Minimization Principle associated with Laplace's Equation

We know from the study of soap films that the Laplace–Young equation is related to the minimization of the free energy associated with the soap films and bubbles. When the excess pressure across the soap film is zero this minimization principle is equivalent to the minimization of the area of the film.

The Poisson equation for the soap film, (6.53), is obtained when the gradient of the soap film is small. So we might expect that the differential equations of Poisson and Laplace are related to a minimization principle.

Let us consider again the electrostatic example of a two dimensional electrostatic field $E(x,y)$, with potential $V(x,y,)$ in a medium of dielectric constant ϵ and zero charge density. The analogue of this situation is the soap film with zero excess pressure and small gradients. One might conjecture that the minimum area property of the soap film, or minimum energy of the soap film, may correspond in the electrostatic problem to the minimization of the electrostatic energy of the field.

The energy of the two dimensional electrostatic field, W, is given by

$$W = \tfrac{1}{2}\epsilon \int \int_S \mathbf{E}^2 \, dx \, dy, \tag{6.67}$$

where the integral is taken over the surface S. Now in terms of the components of $\mathbf{E}$, E_x and E_y,

$$\mathbf{E} = \hat{\mathbf{i}}E_x + \hat{\mathbf{j}}E_y, \tag{6.68}$$

where $\hat{\mathbf{i}}$ and $\hat{\mathbf{j}}$ are the unit vectors in the x and y directions. Also

$$E_x = -\frac{\partial V}{\partial x}, \tag{6.69}$$

and

$$E_y = -\frac{\partial V}{\partial y}. \tag{6.70}$$

Substituting (6.68), (6.69) and (6.70) into (6.67),

$$W = \tfrac{1}{2}\epsilon \int \int_S \left[\left(\frac{\partial V}{\partial x}\right)^2 + \left(\frac{\partial V}{\partial y}\right)^2 \right] dx \, dy. \tag{6.71}$$

The potential, V, that gives rise to the minimum value of W is given by the two dimensional Euler–Lagrange equation of Appendix AIV, Eq. (AIV.9). Applying this equation gives

$$\frac{\partial^2 V}{\partial x^2} + \frac{\partial^2 V}{\partial y^2} = 0. \tag{6.72}$$

So, indeed, Laplace's equation results from the minimization of the electrostatic energy of the system. This result can easily be extended to all problems that require the solution of Laplace's equation, and can be generalized to include Poisson's equation. However this latter analysis is beyond the scope of this book.

The result relating the minimum area and Laplace's differential equation in the case of small gradients could have been observed earlier if we had noticed that the soap film has an area, A, which can be written in Cartesian coordinates, from AIV.11, as

$$A = \int \int (1 + z_x^2 + z_y^2)^{\frac{1}{2}} \, dx \, dy, \tag{6.73}$$

where the integral is over the whole surface and in the usual notation,

$$z_x = \frac{\partial z}{\partial x} \quad \text{and} \quad z_y = \frac{\partial z}{\partial y}. \tag{6.74}$$

This can be expanded for small z_x and z_y, by the binomial theorem, as

$$A = \int \int [1 + \tfrac{1}{2}(z_x^2 + z_y^2) + \ldots] \, dx \, dy. \tag{6.75}$$

The minimization of A occurs, providing we retain only the lowest order terms, when

$$\int \int (z_x^2 + z_y^2) \, dx \, dy \tag{6.76}$$

is minimized. This is of exactly the same form as the electrostatic energy (6.71). Both (6.76) and (6.71) contain integrands that are quadratic in the sum of the first partial derivatives. This form of the energy integral always leads to Laplace's equation. It should be remembered that (6.75) is an approximation for the area, for small z_x and z_y, but (6.71) is an exact relationship for the potential, V, that satisfies Laplace's equation.

6.9   Concluding Remarks

At the end of our 'journey' we can look back and appreciate the contributions made by scientists from all disciplines to the study of soap films and soap bubbles. They have produced many fundamental discoveries which have provided a greater understanding of surface phenomena and led to important technological developments. Today both industrial and public research bodies recognize the importance of work on surface phenomena and support large research programmes throughout the world.

The mathematics of minimum surfaces is also being pursued with vigour in universities and research institutes. Mathematicians have been stimulated by the experiments of Joseph Plateau. In particular Jesse Douglas obtained the highest mathematical award, the Field's Medal, for his work on the *Plateau Problem*. In the future we can look forward to continued investigations into all aspects of surface properties.

The educational value of soap films and bubbles should not be overlooked in these concluding remarks. The shapes, motions, and colours of films and bubbles provide a simple means of demonstrating many interesting phenomena to students at all academic levels, from primary school to university. It is my hope that this volume will encourage greater use of soap film and bubble demonstrations and experiments in our schools, educational institutions, and universities.

APPENDIX I

The Euler–Lagrange Equation

The simplest problems in the calculus of variations require the determination of the extremum value of an integral J, with an integrand that is a function of one independent variable, x, a dependent variable, $y(x)$, and

$$y_x = \frac{dy}{dx}. \tag{AI.1}$$

The limits, x_1 and x_2, of the integral are fixed. If the integrand function is

$$f(x, y, y_x), \tag{AI.2}$$

then

$$J = \int_{x_1}^{x_2} f(x, y, y_x) \, dx. \tag{AI.3}$$

The functional form, $y(x)$, is unknown and must be determined. It is required to find the extremum value of J subject to a variation in $y(x)$, in which the end points, (x_1, y_1) and (x_2, y_2), remain fixed. The varied path $y(x)$ must thus pass through (x_1, y_1) and (x_2, y_2) (Fig. AI.1). This problem is more difficult than the determination of the stationary points of a function.

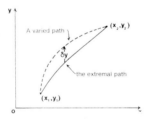

Fig. AI.1   The extremum path, full curve, and a varied path, broken curve.

In order to obtain an equation for the extremum path it is necessary to introduce a parameter, a, that will enable the path to be varied, and an arbitrary function, $\eta(x)$, that vanishes at the end points. That is

$$\eta(x_1) = \eta(x_2) = 0. \tag{AI.4}$$

It is now possible to introduce a varied path, $y(x, a)$, which has the properties that it passes through (x_1, y_1) and (x_2, y_2) for all values of a and gives the extremum path, $y(x, 0)$, when $a = 0$. The varied path may be written

$$y(x, a) = y(x, 0) + a\eta(x). \qquad \text{(AI.5)}$$

A varied path is shown in Fig. (AI.1) by the broken line. For any value of a the path $y(x, a)$ passes through the end points, from definition (AI.4), and gives the extremum path $y(x, 0)$ when $a = 0$.

Substituting (AI.5) into (AI.3), the value of J for any varied path with parameter a, $J(a)$, is given by

$$J(a) = \int_{x_1}^{x_2} f(x, y(x, a), y_x(x, a)) \, dx. \qquad \text{(AI.6)}$$

The requirement that $J(a)$ be an extremum gives

$$\left(\frac{\partial J(a)}{\partial a} \right)_{a=0} = 0. \qquad \text{(AI.7)}$$

Differentiating (AI.6) with respect to a,

$$\frac{\partial J(a)}{\partial a} = \int_{x_1}^{x_2} \left(\frac{\partial f}{\partial y} \frac{\partial y}{\partial a} + \frac{\partial f}{\partial y_x} \frac{\partial y_x}{\partial a} \right) dx. \qquad \text{(AI.8)}$$

From (AI.5),

$$\frac{\partial y}{\partial a} = \eta(x), \qquad \text{(AI.9)}$$

and

$$\frac{\partial y_x}{\partial a} = \frac{d\eta(x)}{dx}. \qquad \text{(AI.10)}$$

It is necessary to assume that $y(x, a)$ and $y_x(x, a)$ are both continuous in order to perform integral (AI.8). Equation (AI.8) becomes, on substituting (AI.9) and (AI.10),

$$\frac{\partial J(a)}{\partial a} = \int_{x_1}^{x_2} \left(\frac{\partial f}{\partial y} \eta(x) + \frac{\partial f}{\partial y_x} \frac{\partial \eta}{dx} \right) dx. \tag{AI.11}$$

The second term on the right hand side of (AI.11) can be integrated by parts,

$$\int_{x_1}^{x_2} \frac{\partial f}{\partial y_x} \frac{d\eta}{dx} dx = \left[\frac{\partial f}{\partial y_x} \eta(x) \right]_{x_1}^{x_2} - \int_{x_1}^{x_2} \eta(x) \frac{d}{dx} \left(\frac{\partial f}{\partial y_x} \right) dx. \tag{AI.12}$$

Now $\eta(x)$ is zero at x_1 and x_2, from (AI.4), so the first term on the right hand side of (AI.12) is zero, leaving

$$\int_{x_1}^{x_2} \frac{\partial f}{\partial y_x} \frac{d\eta}{dx} dx = - \int_{x_1}^{x_2} \eta(x) \frac{d}{dx} \left(\frac{\partial f}{\partial y_x} \right) dx. \tag{AI.13}$$

Substituting (AI.13) back into (AI.11),

$$\frac{\partial J(a)}{\partial a} = \int_{x_1}^{x_2} \left[\frac{\partial f}{\partial y} - \frac{d}{dx} \left(\frac{\partial f}{\partial y_x} \right) \right] \eta(x) \, dx. \tag{AI.14}$$

The extremum value of $J(a)$ occurs when $\partial J(a)/\partial a = 0$ and we have chosen it to occur when $a = 0$, (AI.5). Thus for the extremum value of $J(a)$ (AI.14) becomes

$$0 = \int_{x_1}^{x_2} \left[\frac{\partial f}{\partial y} - \frac{d}{dx} \left(\frac{\partial f}{\partial y_x} \right) \right] \eta(x) \, dx, \tag{AI.15}$$

where y is now $y(x,0)$, the extremum path. This equation is valid for any arbitrary function, $\eta(x)$, satisfying (AI.4). Consequently (AI.15) can only be satisfied if

$$\frac{\partial f}{\partial y} - \frac{d}{dx} \left(\frac{\partial f}{\partial y_x} \right) = 0. \tag{AI.16}$$

This is the Euler–Lagrange Equation.

An Alternative Form of the Euler–Lagrange Equation

An alternative form of the Euler–Lagrange equation, (AI.16), is convenient when f does not depend explicitly on x. Consider multiplying (AI.16) by y_x to give

$$y_x \left[\frac{\partial f}{\partial y} - \frac{d}{dx}\left(\frac{\partial f}{\partial y_x} \right) \right] = 0. \qquad (AI.17)$$

Now differentiating $f(x, y, y_x)$,

$$\frac{df}{dx} = \frac{\partial f}{\partial y} y_x + \frac{\partial f}{\partial y_x} y_{xx} + \frac{\partial f}{\partial x} , \qquad (AI.18)$$

where $y_{xx} = d^2y/dx^2$. Hence

$$y_x \frac{\partial f}{\partial y} = \frac{df}{dx} - \frac{\partial f}{\partial y_x} y_{xx} - \frac{\partial f}{\partial x} . \qquad (AI.19)$$

Substituting from (AI.19) into (AI.17)

$$\frac{df}{dx} - \frac{\partial f}{\partial y_x} y_{xx} - \frac{\partial f}{\partial x} - y_x \frac{d}{dx}\left(\frac{\partial f}{\partial y_x} \right) = 0. \qquad (AI.20)$$

Simplifying this equation gives,

$$-\frac{\partial f}{\partial x} + \frac{d}{dx}\left(f - y_x \frac{\partial f}{\partial y_x} \right) = 0. \qquad (AI.21)$$

This is the alternative form of the Euler–Lagrange equation. It is particularly useful when f does not depend explicitly on x, so that

$$\frac{\partial f}{\partial x} = 0. \qquad (AI.22)$$

Then (AI.21) becomes

$$\frac{d}{dx}\left(f - y_x \frac{\partial f}{\partial y_x} \right) = 0. \qquad (AI.23)$$

This can be integrated to give

$$f - y_x \frac{\partial f}{\partial y_x} = c, \qquad (AI.24)$$

where c is a constant.

APPENDIX II

The Shortest Path Joining Two Points

The shortest path joining two points is the simplest example of the application of the Euler–Lagrange equation derived in Appendix I.

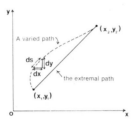

Fig. AII.1   The shortest path linking (x_1, y_1) and (x_2, y_2), full curve, and a varied path, broken curve.

Let the two points have Cartesian coordinates (x_1, y_1) and (x_2, y_2). The element of distance, ds, along a varied path, such as that shown in Fig. AII.1, is given by Pythagoras's theorem as

$$ds = [(dx)^2 + (dy)^2]^{1/2}. \qquad (AII.1)$$

The total length of the varied path joining (x_1, y_1) and (x_2, y_2) is

$$s = \int_{x_1}^{x_2} [(dx)^2 + (dy)^2]^{1/2}, \qquad (AII.2)$$

$$= \int_{x_1}^{x_2} \left[1 + \left(\frac{dy}{dx}\right)^2\right]^{1/2} dx. \qquad (AII.3)$$

We require the extremum value of s. The integrand in (AII.3) is of the form required for the application of the Euler–Lagrange equation in which the function $f(x, y, y_x)$, of (AI.3), is given by

$$f = (1 + y_x^2)^{1/2}. \qquad (AII.4)$$

160

Applying the standard form of the Euler–Lagrange equation (AI.16), the extremum path is given by

$$\frac{\partial f}{\partial y} - \frac{d}{dx}\left(\frac{\partial f}{\partial y_x}\right) = 0. \tag{AII.5}$$

From (AII.4) the term in $\partial f/\partial y$ is absent, so (AII.5) becomes

$$\frac{d}{dx}\left(\frac{\partial f}{\partial y_x}\right) = 0. \tag{AII.6}$$

Integrating (AII.6),

$$\frac{\partial f}{\partial y_x} = K, \tag{AII.7}$$

where K is a constant. Substituting into (AII.7) for f, from (AII.4),

$$\frac{\partial}{\partial y_x}(1+y_x^2)^{1/2} = K. \tag{AII.8}$$

Differentiating the left hand side,

$$y_x(1+y_x^2)^{-1/2} = K. \tag{AII.9}$$

The left hand side of this expression is a function of y_x only. So y_x must be a constant. Let

$$y_x = m, \tag{AII.10}$$

where m is a constant. It will be related to K by (AII.9). That is

$$m(1+m^2)^{-1/2} = K. \tag{AII.11}$$

From (AII.10),

$$\frac{dy}{dx} = m. \tag{AII.12}$$

Integrating (AII.12),

$$y = mx+c, \tag{AII.13}$$

where c is a constant. This is a straight line. It must pass through the two end points (x_1, y_1) and (x_2, y_2), by the assumptions made in the derivation of the Euler–Lagrange equation, consequently the constants m and c can be expressed in terms of (x_1, y_1) and (x_2, y_2). So the extremum path is a straight line passing through the two end points. It requires further analysis to prove that it is the path of minimum length. However by examination of the varied paths close to the extremum, or on physical grounds, it can be shown to be the path of minimum length.

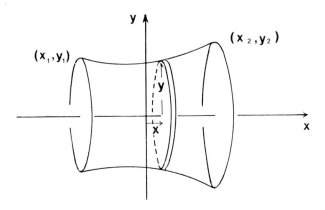

Fig. AIII.1 An axially symmetric surface joining two coaxial rings which are perpendicular to their common axis.

APPENDIX III

The Minimum Surface Area Bounded by Two Coaxial Rings

The minimum surface area bounded by two coaxial rings perpendicular to the axis will, by symmetry, be symmetrical about the axis. Thus it is sensible to consider a sequence of varied surfaces that are symmetric about the axis and are bounded by the two rings (Fig. AIII.1), when applying the Euler–Lagrange equation. The axis of symmetry is taken as the x-axis. In the x–y plane an element of the length of the surface is ds (Fig. AIII.1). The area of surface contained between two planes perpendicular to the axis at x and $x+dx$ is

$$dA = 2\pi y ds. \qquad \text{(AIII.1)}$$

Now expressing ds in Cartesian coordinates,

$$ds = (1+y_x^2)^{1/2}\, dx, \qquad \text{(AIII.2)}$$

where

$$y_x = \frac{dy}{dx}. \qquad \text{(AIII.3)}$$

Therefore

$$dA = 2\pi y(1+y_x^2)^{1/2}dx. \qquad \text{(AIII.4)}$$

If the rings are at positions x_1 and x_2, the total area of the surface is

$$A = \int_{x_1}^{x_2} 2\pi y(1+y_x^2)^{1/2}\, dx. \qquad \text{(AIII.5)}$$

Thus A is in the appropriate form for the application of the Euler–Lagrange equation. The Euler–Lagrange function, f, Eq. (AI.16), is given by

$$f = 2\pi y(1+y_x^2)^{1/2}. \qquad \text{(AIII.6)}$$

163

Applying the alternative form of the Euler–Lagrange equation, as f does not depend explicitly on x, Eq. (AI.23),

$$\frac{d}{dx}\left(f-y_x\frac{\partial f}{\partial y_x}\right) = 0. \tag{AIII.7}$$

Integrating this equation,

$$f-y_x\frac{\partial f}{\partial y_x} = c, \tag{AIII.8}$$

where c is a constant. Substituting for f from (AIII.6),

$$y(1+y_x^2)^{1/2}-y_x\frac{\partial}{\partial y_x}[y(1+y_x^2)^{1/2}] = c_1, \tag{AIII.9}$$

where $c_1 = c/2\pi$.
Performing the differentiation in (AIII.9),

$$y(1+y_x^2)^{1/2}-y_x^2y(1+y_x^2)^{-1/2} = c_1. \tag{AIII.10}$$

Simplifying (AIII.10),

$$y(1+y_x^2)^{-1/2} = c_1. \tag{AIII.11}$$

This gives

$$\frac{dx}{dy} = c_1(y^2-c_1^2)^{-1/2}. \tag{AIII.12}$$

Integrating,

$$x = c_1 \cosh^{-1}\left(\frac{y}{c_1}\right)+c_2, \tag{AIII.13}$$

where c_2 is the integration constant. Reverting (AIII.13),

$$y = c_1 \cosh\left(\frac{x-c_2}{c_1}\right). \tag{AIII.14}$$

This is a catenary of revolution about the x-axis, or catenoid, bounded by the two rings. There is not necessarily a unique solution to (AIII.14) and further investigation of this solution is necessary. Consider the possible solutions for $y(x)$ that pass through (x_1, y_1). Figure (AIII.2) shows some of these catenaries.

The broken curve, OT, is the tangent curve to the system of catenaries passing through (x_1, y_1). Only one catenary passes through a point, such as M, on the broken curve OT. A point such as B to the left of the curve OT has two catenaries passing through it. No catenary solution exists at a point, such as Z, to the right of curve OT.

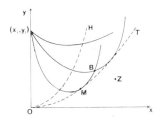

Fig. AIII.2   The broken curve, OT, is the envelope of a sequence of catenaries passing through (x_1, y_1). The full curves are the catenaries.

Bliss[†] has shown, in a detailed account of the solution to this problem, that there always exists a solution consisting of two plane discs bounded by the rings with no intermediate surface. This is known as the Goldschmidt[92] discontinuous solution after the man who first proved its existence in 1831. If (x_2, y_2) lies on, or to the right, of the broken curve OT (Fig. AIII.2), the Goldschmidt solution is the only minimum surface. When (x_2, y_2) lies to the left of the curve OT there are two catenaries that pass through (x_2, y_2). However no minimum curve is generated by the catenary with its point of tangency to OT in the interval where

$$x_1 < x < x_2. \tag{AIII.15}$$

The broken curve OH (Fig. AIII.2) is the locus of the point (x_2, y_2) for which the area of the Goldschmidt discontinuous surface is equal to the surface area generated by the catenary that does *not* have a point of tangency with OT in the interval given by (AIII.15). When (x_2, y_2) lies between the two broken curves, OH and OT, the Goldschmidt solution gives the surface with the absolute minimum area and the catenoid surface is a relative minimum. When (x_2, y_2) lies to the left of the curve OH the catenary gives the absolute minimum area and the Goldschmidt solution gives the relative minimum.

[†] For a detailed analysis of this problem reference should be made to *Calculus of Variations*, G. A. Bliss, Chapter IV, Open Court Pub. Co., Illinois, U.S.A. (1944).

It is instructive to examine a specific example. Consider two rings of equal radius y_0 symmetrically positioned about the y-axis at $\pm x_0$ then

$$(x_1, y_1) = (x_0, y_0), \tag{AIII.16}$$

and

$$(x_2, y_2) = (-x_0, y_0). \tag{AIII.17}$$

By symmetry (AIII.14) requires

$$c_2 = 0, \tag{AIII.18}$$

and substituting (AIII.16) into (AIII.14),

$$\lambda y_0 = \cosh(\lambda x_0), \tag{AIII.19}$$

where

$$\lambda = 1/c_1. \tag{AIII.20}$$

Now consider the variation of the catenary with x_0. x_0 will have its maximum value, x_{0m}, with respect to a change in λ when

$$\frac{dx_0}{d\lambda} = 0. \tag{AIII.21}$$

Differentiating (AIII.19) with respect to λ, keeping y_0 fixed,

$$y_0 = \left(x_0 + \lambda \frac{dx_0}{d\lambda}\right) \sinh(\lambda x_0). \tag{AIII.22}$$

x_{0m} is given by condition (AIII.21). Hence

$$y_0 = x_{0m} \sinh(\lambda x_{0m}). \tag{AIII.23}$$

Substituting for y_0 from AIII.19,

$$\lambda x_{0m} = \coth(\lambda x_{0m}). \tag{AIII.24}$$

The numerical solution to this equation gives

$$\lambda x_{0m} = 1.200. \tag{AIII.25}$$

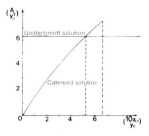

Fig. AIII.3 The area, A, of the catenoid surface and the Goldschmidt solution as a function of separation, x, of the two rings of equal radius, y_0.

Substituting this result into (AIII.19),

$$x_{0m} = 0.663 y_0. \qquad (AIII.26)$$

Figure AIII.3 shows the variation of the area of the minimum area catenoid as x_0 is varied. It gives an absolute or local minimum. The Goldschmidt discontinuous solution, consisting of two discs of radius y_0, is the absolute minimum for

$$\frac{x_0}{y_0} > 0.528. \qquad (AIII.27)$$

The catenary of revolution is the absolute minimum for

$$\frac{x_0}{y_0} < 0.528. \qquad (AIII.28)$$

For

$$0.528 < \frac{x_0}{y_0} < 0.663 \qquad (AIII.29)$$

the catenoid is only a local minimum.

APPENDIX IV

The Euler–Lagrange Equation for Two Independent Variables, Minimum Areas, and the Laplace–Young Equation

In order to apply the Euler–Lagrange formalism to the problem of determining the minimum area contained by a simple closed curve it is necessary to generalize the result of Appendix I to the case of two independent variables.

It is required to determine the extremum value of the two dimensional integral

$$J = \iint f(x, y, u, u_x, u_y)\,dx\,dy, \qquad (AIV.1)$$

subject to a variation in the unknown function $u(x,y)$. The functional form of $f(x,y,u,u_x,u_y)$ is known. We shall continue to use the notation u_x and u_y for the partial derivative of u with respect to x and y.

Generalizing the procedure of Appendix I we introduce a varied function $u(x,y,a)$ which gives the extremum value of J when $a = 0$. The function $u(x,y,a)$ is defined by

$$u(x, y, a) = u(x,y,0) + a\eta(x,y), \qquad (AIV.2)$$

where $\eta(x,y)$ is an arbitrary function which vanishes at the boundary and is differentiable. From (AIV.2)

$$u_x(x, y, a) = u_x(x,y,0) + a\eta_x, \qquad (AIV.3)$$

and

$$u_y(x, y, a) = u_y(x,y,0) + a\eta_y. \qquad (AIV.4)$$

Differentiating (AIV.1) for the varied functions,

$$\frac{\partial J(a)}{\partial a} = \iint \left(\frac{\partial f}{\partial u}\eta + \frac{\partial f}{\partial u_x}\eta_x + \frac{\partial f}{\partial u_y}\eta_y \right) dx\,dy. \qquad (AIV.5)$$

Integrating the terms $(\partial f/\partial u_x)\eta_x$ and $(\partial f/\partial u_y)\eta_y$ by parts, as in (AI.12) and (AI.13), (AIV.5) becomes,

$$\frac{\partial J(a)}{\partial a} = \iint \left(\frac{\partial f}{\partial u} - \frac{\partial}{\partial x}\frac{\partial f}{\partial u_x} - \frac{\partial}{\partial y}\frac{\partial f}{\partial u_y} \right)\eta(x,y)\,dx\,dy. \qquad (AIV.6)$$

The extremum value of J occurs when

$$\frac{\partial J(\alpha)}{\partial \alpha} = 0. \tag{AIV.7}$$

That is, from AIV.6,

$$0 = \int \int \left(\frac{\partial f}{\partial u} - \frac{\partial}{\partial x} \frac{\partial f}{\partial u_x} - \frac{\partial}{\partial y} \frac{\partial f}{\partial u_y} \right) \eta(x,y) \, dx \, dy. \tag{AIV.8}$$

Since $\eta(x,y)$ is arbitrary (AIV.8) will only be satisfied if

$$\frac{\partial f}{\partial u} - \frac{\partial}{\partial x} \frac{\partial f}{\partial u_x} - \frac{\partial}{\partial y} \frac{\partial f}{\partial u_y} = 0. \tag{AIV.9}$$

This is the generalized two dimensional form of the Euler–Lagrange equation.

Application to the Minimum Area Problem

A simple closed curve Γ in space, with a projection on the $z = 0$ plane that forms a simple closed curve, has surfaces bounded by Γ that can be written in the form

$$z = u(x,y), \tag{AIV.10}$$

where $u(x,y)$ is a function that satisfies the boundary curve Γ.

The surface area, by a generalization of (AII.3), is

$$A = \int \int_\Gamma (1 + u_x^2 + u_y^2)^{1/2} \, dx \, dy. \tag{AIV.11}$$

The integral is bounded by the closed curve Γ. The condition for a minimum in A is, from (AIV.9),

$$\frac{\partial}{\partial x} \frac{\partial f}{\partial u_x} + \frac{\partial}{\partial y} \frac{\partial f}{\partial u_y} = 0, \tag{AIV.12}$$

where

$$f = (1 + u_x^2 + u_y^2)^{1/2}. \tag{AIV.13}$$

It has become conventional to use the following notation,

$$p = u_x, \qquad q = u_y, \qquad \rho = 1+p^2+q^2, \qquad \text{(AIV.14)}$$

$$r = u_{xx}, \qquad s = u_{xy}, \qquad \text{and} \qquad t = u_{yy}. \qquad \text{(AIV.15)}$$

So from (AIV.12–15) the condition for a minimum value of A is

$$\frac{\partial}{\partial x}\frac{p}{\rho} + \frac{\partial}{\partial y}\frac{q}{\rho} = 0. \qquad \text{(AIV.16)}$$

Differentiating this result gives,

$$r(1+q^2) - 2pqs + t(1+p^2) = 0. \qquad \text{(AIV.17)}$$

This result was first obtained by Lagrange.[101] However Meusnier[106] re-expressed the equation in terms of the radii of curvature at any point on the minimum surface, ρ_1 and ρ_2, formed by any orthogonal pair of axes in the surface and obtained the Laplace–Young result in the case of zero excess pressure (see Eq. 5.10),

$$\frac{1}{\rho_1} + \frac{1}{\rho_2} = 0. \qquad \text{(AIV.18)}$$

In terms of the principal radii of curvature R_1 and R_2 this gives

$$\frac{1}{R_1} + \frac{1}{R_2} = 0. \qquad \text{(AIV.19)}$$

The assumptions made about Γ and the enclosed surfaces are too restrictive to cover all surfaces of minimum area contained by all closed boundaries. The analysis however illustrates the point that the Laplace–Young equation can be deduced from the minimum area property.

APPENDIX V

The Maximum Area Contained by a Given Circumference

The solution to the problem of determining the maximum area contained by a fixed circumference has been solved using the minimum area property of soap films in Chapter 1. The solution was found to be the area of the disc contained by the circular circumference.

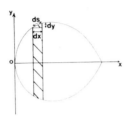

Fig. AV.1 An element of area of a closed curve of fixed length that passes through the origin and is symmetrical about the x-axis.

In order to solve this problem analytically, using the calculus of variations, consider a closed curve passing through the origin of a Cartesian system of coordinates. Let us assume that the curve is symmetric about the x-axis (Fig. AV.1). This assumption is restrictive, but will simplify the analysis.

An element of area dA of width dx is given by,

$$dA = 2y\,dx. \tag{AV.1}$$

If the length of the circumference is $2l$ and an element dx is associated with an element of length ds (Fig. AV.1), then from (AV.1) and Pythagoras's theorem,

$$dA = 2y[(ds)^2 - (dy)^2]^{1/2}, \tag{AV.2}$$

$$= 2y(1 - y_s^2)^{1/2}\,ds, \tag{AV.3}$$

where

$$y_s = \frac{dy}{ds}. \tag{AV.4}$$

171

The total area is

$$A = 2 \int_0^l y(1 - y_s^2)^{1/2} \, ds. \tag{AV.5}$$

Using the alternative form of the Euler–Lagrange equation, (AI.24), the curve $y(x)$ with the maximum area is given by

$$y(1 - y_s^2)^{1/2} + yy_s^2(1 - y_s^2)^{-1/2} = c, \tag{AV.6}$$

where c is a constant. Simplifying (AV.6),

$$cy_s = \pm(c^2 - y^2)^{1/2}. \tag{AV.7}$$

That is , using the positive sign,

$$\int_0^y \frac{c\,dy}{(c^2 - y^2)^{1/2}} = \int_0^s ds. \tag{AV.8}$$

Integrating,

$$c \sin^{-1}\left(\frac{y}{c}\right) = s. \tag{AV.9}$$

Reverting,

$$y = c \sin\left(\frac{s}{c}\right). \tag{AV.10}$$

As $y = 0$ when $s = l$, (AV.10) gives

$$c = l/\pi. \tag{AV.11}$$

Thus (AV.10) becomes,

$$y = \frac{l}{\pi} \sin \frac{\pi s}{l}. \tag{AV.12}$$

In order to obtain the equation $y(x)$ of this curve differentiate (AV.12) to obtain,

$$dy = \cos\left(\frac{\pi s}{l}\right) ds, \tag{AV.13}$$

and use the result

$$(dx)^2 = (ds)^2 - (dy)^2,$$ (AV.14)

to obtain dx. Substituting (AV.13) into (AV.14),

$$dx = \sin\left(\frac{\pi s}{l}\right) ds.$$ (AV.15)

Integrating from $x = 0$, at $s = 0$, to (x, s) gives

$$x = \frac{l}{\pi} - \frac{l}{\pi} \cos\left(\frac{\pi s}{l}\right).$$ (AV.16)

Thus from (AV.16) and (AV.12)

$$\left(x - \frac{l}{\pi}\right)^2 + y^2 = \frac{l^2}{\pi^2}.$$ (AV.17)

This is a circle of radius (l/π), which will have an area of (l^2/π). Thus the maximum area contained by a fixed perimeter occurs when the perimeter forms the circumference of a circle.

APPENDIX VI

The Maximum Volume Contained by a Closed Surface of Fixed Area

It is well known that the maximum volume contained by a closed surface of given area occurs when the surface is that of a sphere. This result can be proved generally. Here we shall use the calculus of variations to prove it under the restriction that the surface is a surface of revolution, about the x-axis, passing through the origin.

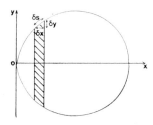

Fig. AVI.1 A volume element of a closed surface of fixed area that passes through the origin and is symmetrical about the x-axis.

The volume, V, of a surface of revolution (Fig. AVI.1), is given in Cartesian coordinates by

$$V = \pi \int y^2 dx. \qquad (AVI.1)$$

This volume is subject to the condition that its area, A, is fixed. That is

$$A = 2\pi \int y ds, \qquad (AVI.2)$$

where ds is a line element of the surface in the x–y plane.

We cannot apply the Euler–Lagrange equation directly to (AVI.1) as there is the constraint imposed by (AVI.2). However we can make a transformation of variables so that condition (AVI.2) is incorporated in the limits of integration of (AVI.1). Let us make a change of variable so that

$$y ds = d\omega. \qquad (AVI.3)$$

174

From Eq. (AVI.2), ω can vary from 0 to the fixed value $(A/2\pi)$. Let us put $a = A/2\pi$. Now as

$$(dx)^2+(dy)^2 = (ds)^2, \tag{AVI.4}$$

on dividing by $(d\omega)^2$ we obtain

$$\left(\frac{dx}{d\omega}\right)^2 + \left(\frac{dy}{d\omega}\right)^2 = \left(\frac{ds}{d\omega}\right)^2. \tag{AVI.5}$$

Using the notation where subscripts label the derivatives this becomes, from (AVI.3),

$$x_\omega^2 + y_\omega^2 = \frac{1}{y^2}. \tag{AVI.6}$$

V can now be written, from (AVI.1), in terms of ω as

$$V = \pi \int_0^a y^2 \frac{dx}{d\omega} d\omega, \tag{AVI.7}$$

$$= \pi \int_0^a y^2 x_\omega d\omega. \tag{AVI.8}$$

Substituting for x_ω from (AVI.6),

$$V = \pi \int_0^a y(1-y^2 y_\omega^2)^{1/2} d\omega. \tag{AVI.9}$$

We have now converted the problem into one with fixed end points. Before applying the Euler–Lagrange equation it is convenient to make a further transformation by substituting into (AVI.9)

$$z = \tfrac{1}{2} y^2. \tag{AVI.10}$$

This gives

$$V = \pi\sqrt{2} \int_0^a [z(1-z_\omega^2)]^{1/2} d\omega. \tag{AVI.11}$$

Applying the alternative form of the Euler–Lagrange equation, (AI.24), as the integrand does not depend explicitly on ω,

$$z = b(1 - z_w{}^2), \qquad (AVI.12)$$

where b is a constant. Equation (AVI.12) can be rewritten as

$$\frac{dz}{d\omega} = \left(1 - \frac{z}{b}\right)^{1/2}. \qquad (AVI.13)$$

So

$$\int\limits_0^z \left(1 - \frac{z}{b}\right)^{-1/2} dz = \int\limits_0^\omega d\omega, \qquad (AVI.14)$$

as $z = 0$ when $\omega = 0$. Integrating (AVI.14),

$$-2b\left(1 - \frac{z}{b}\right)^{1/2} + 2b = \omega. \qquad (AVI.15)$$

Therefore

$$4b^2\left(1 - \frac{z}{b}\right) = (\omega - 2b)^2. \qquad (AVI.16)$$

As $z = 0$ for $\omega = a$,

$$b = \frac{a}{4}. \qquad (AVI.17)$$

Thus (AVI.16) becomes, after substituting for b,

$$z = \omega - \frac{\omega^2}{a}. \qquad (AVI.18)$$

This relation gives the maximizing curve.

Now let us return to the original coordinate system. From (AVI.10) and (AVI.6)

$$x_\omega{}^2 = \frac{1}{2z}(1 - z_\omega{}^2). \qquad (AVI.19)$$

Substituting for $z_w{}^2$ from (AVI.18),

$$x_\omega{}^2 = \frac{2}{a}. \qquad (AVI.20)$$

Integrating gives,

$$x = \left(\frac{2}{a}\right)^{1/2} \omega. \qquad \text{(AVI.21)}$$

Eliminating ω from (AVI.18), (AVI.10) and (AVI.21) gives the maximizing surface,

$$\left[x - \left(\frac{a}{2}\right)^{1/2}\right]^2 + y^2 = \left(\frac{a}{2}\right). \qquad \text{(AVI.22)}$$

Thus the surface of revolution is a sphere of radius $(a/2)^{1/2}$, and the maximizing volume is $(4/3)\pi(a/2)^{3/2}$. This is the sphere with surface area A as $a = A/2\pi$

REFERENCES

I Popular texts

1. Boys, C. V. (1890) *Soap Bubbles and the forces which mould them* (Society for the Promotion of Christian Knowledge, London; E. and J. B. Young, New York; new and enlarged edition, C.P.C.K., London, F. S. Graham, New York 1912; many other editions including that by William Heinemann, London 1965, and the enlarged edition, *Soap Bubbles their colours and the forces which mould them*, Dover, New York 1959).
2. Stevens, P. S. (1976) *Patterns of Nature* (Penguin).

II General texts

3. Bikerman, J. J. (1970) *Physical Surfaces* (Academic).
4. Bikerman, J. J. (1973) *Foams* (Springer–Verlag).
5. Wolf, K. L. (1968) *Tropfen, Blasen und Lamellen* (Springer–Verlag).
6. Adam, N. K. (1941—3rd ed.) *The Physics and Chemistry of Surfaces* (Oxford).
7. Overbeek, J. Th. G. (Ed.) (1967) *Chemistry, Physics and Application of Surface Active Substances* (Gordon and Breach), pp. 19–37.
8. Lawrence, A. S. C. (1929) *Soap Films, a study of molecular individuality* (Bell).
9. Adamson, A. W. (1967—2nd ed.) *Physical Chemistry of Surfaces* (Interscience).
10. Harkins, W. D. (1952) *The Physical Chemistry of Surface Films* (Reinhold).
11. Wolf, K. L. (Vol. I—1957, Vol. II—1959) *Physik und Chemie der Grenzflächen* (Springer–Verlag).
12. Smith, C. J. (1962—2nd ed.) *A Degree Physics, Part I—The General Properties of Matter* (Edward Arnold), Ch. X.
13. Champion, F. C. and Davy, N. (1961) *Properties of Matter* (Blackie), Ch. VII.

III Mathematical texts

14. Courant, R. and Robbins, H. (1973) *What is Mathematics?* (Oxford), pp. 329–361 and 385–397.
15. Newman, J. R. (1956) *The World of Mathematics Vol. II* (Simon and Schuster), pp. 882–909.
16. Pars, L. A. (1962) *An Introduction to the Calculus of Variations* (Heinemann).
17. Arfken, G. (1970) *Mathematical Methods for Physicists* (Academic), Ch. 17.
18. Bliss G. A. (1925) *Calculus of Variations* (Open Court).
19. Coxeter, H. S. M. and Greitzer, S. L. (1967) *Geometry Revisited* (Random House).
20. Lowry, H. V. and Hayden, H. A. (1955) *Advanced Mathematics for Technical Students, Part II* (Longmans, Green), p. 255.
21. Nitsche, J. C. C. (1975) *Vorlesungen über Minimalflächen* (Springer–Verlag).
22. Cundy, H. M. and Rollett, A. P. (1961—2nd ed.) *Mathematical Models* (Clarendon).
23. Schwarz, H. A. (1890) *Gesammelte Mathematische Abhandlungen*, 2 Vols. (Springer).
24. Radó, T. (1951) *On the Problem of Plateau* (Chelsea).
25. Almgren, F. J. Jr. (1966) *Plateau's Problem, An Invitation to Varifold Geometry* (Benjamin).
26. Christofides, N. (1975) *Graph Theory. An Algorithmic Approach* (Academic), pp. 142–145.

IV Specialist texts

27. Cereijido, M. and Rotunno, C. A. (1970) *Introduction to the Study of Biological Membranes* (Gordon and Breach).
28. Defay, R., Prigogine, I., Bellmans, A. and Everett, D. H. (1966) *Surface Tension and Adsorption* (Longmans, Green).
29. Meiners, H. F. (Ed.) (1970) *Physics Demonstration Experiments Vol. I* (The Ronald Press), pp. 461–464.
30. Coulson, C. A. (1955) *Waves* (Oliver and Boyd), Chs. III and V.
31. Rayleigh, Lord (1929) *The Theory of Sound, Vol. I* (Macmillan), Ch. IX.
32. Born, M. and Wolf, E. (1959) *Principles of Optics* (Pergamon).
33. Vašiček, A. (1960) *Optics of Thin Films* (North Holland).
34. Mysels, K. J., Shinoda, K. and Frankel, S. (1959) *Soap Films, Studies of their Thinning and a Bibliography* (Pergamon).
35. Matijevic, E. (Ed.) (1971) *Surface and Colloid Science Vol. 3* (Wiley), pp. 167–239.
36. Akers, R. J. (Ed.) (1977) *Foams* (Academic).
37. Thompson, D'Arcy W. (1961—abridged ed.) *On Growth and Form* (Cambridge), Ch. V.
38. Hair, M. L. (1971) *The Chemistry of Biosurfaces* (Marcel Dekker).
39. Karplus, W. J. and Soroka, W. W. (1959—2nd Ed.) *Analog Methods in Computation and Simulation* (McGraw–Hill).
40. Otto, F. (Ed.) (1973) *Tensile Structures* (M.I.T.).

V Historical texts

41. Euler, L. (1952) *Opera Omnia* (Orell Füssli), (1) Vols. 24–25.
42. Newton, I. (1952) *Opticks* (Dover), Book I, pt. II, exp. 4, and Book II, pt. I, obs. 17–20.
43. Young, T. (1807) *A Course of Lectures on Natural Philosophy and the Mechanical Arts Vol. I* (J. Johnson), pp. 468–469, plate 30, Fig. 448.
44. Laplace, P. S. (1806) *Mécanique céleste* (Impr. Imperiale), Suppl. to 10th book. Also see Laplace's Oeuvres and the English, annotated, translation by N. Bowdich, vol. IV (Little and Brown, 1839).
45. Lagrange, J. L. (1889—4th ed.) *Mécanique analytique* (Gauthier–Villars).
46. Plateau, J. A. F. (1873) *Statique expérimentale et théorique des liquides soumis aux seules forces moléculaires* (Gauthier–Villars, Trubner et cie, F. Clemm), 2 Vols.
47. Taylor, R. (Ed.) (1966) *Scientific Memoirs, The Sources of Science No. 7* (Johnson Reprint Corp.), Vol. 4, pp. 16–43 and Vol. 5, pp. 584–712.
48. Gillispie, C. C. (Ed.) (1975) *Dictionary of Scientific Biography: J. A. F. Plateau, Vol. XI* (Charles Scribner), p. 20.
49. Gibbs, J. W. (1961) *The Scientific Papers of J. Willard Gibbs, Vols I and II* (Dover).
50. Donnan, F. G. and Hass, A. (Eds.) (1936) *A Commentary on the Scientific Writings of J. Willard Gibbs, Vols I and II* (Yale University Press).
51. Wheeler, L. P. (1952—revised ed.) *Josiah Willard Gibbs* (Yale University Press), p. 260.
52. Gibbs, J. W. (1931) *The Collected Works of J. W. Gibbs* (Longmans, Green), p. 55 ff.
53. Dewar, Lady (Ed.) (1927) *Collected Papers of Sir James Dewar* (Cambridge), 2 Vols.
54. Rayleigh, Lord (1964) *Scientific Papers by Lord Rayleigh, Vols I and II* (Dover), Vol. I, pp. 361–401; Vol. II, pp. 103–117.
55. Kline, M. (1972) *Mathematical Thought from Ancient to Modern Times* (Oxford).
56. Linklater, E. (1972) *The Voyage of the Challenger* (John Murray).

VI *Films*

57. *Kubic Bubbles*, Science Feature Film 68, 5 min., colour, The Central Office of Information, Hercules Road, Westminster Bridge Road, London, S.E.1.
58. *Coalescence of Liquid Drops*, 27 min., Higher Educational Film Library, The Scottish Central Film Library, Glasgow G3 7XN.
59. *Bubble Growth*, National Science Film Library of the Canadian Film Institute, Ottawa, Ontario, Canada.
60. *Oil, Soap and Detergent*, BBC/TV Nature of Things Series, 15 min., Scottish Central Film Library, Glasgow G3 7XN.
61. *Surface Tension in Fluid Mechanics*, 29 min., Scottish Central Film Library, Glasgow G3 7XN.
62. *Surfaces, Part 5 of the Properties of Matter Series* by Sir Lawrence Bragg, 18 min., Scottish Central Film Library, Glasgow G3 7XN.
63. *Bubble Model of a Crystal: Structure and Boundaries*, 2 min. 5 sec., 8 mm Film-loop, Ealing Scientific, Watford WD2 4PW.
64. *Bubble Model of a Crystal: Deformation and Dislocations*, 3 min., 8 mm Film-loop, Ealing Scientific, Watford WD2 4PW.
65. *The Surface of Water*, 3 min. 15 sec., 8 mm Film-loop, Ealing Scientific, Watford WD2 4PW.
66. *Drops and Splashes*, 3 min. 30 sec., 8 mm Film-loop, Ealing Scientific, Watford WD2 4PW.
67. *Soap Film Oscillations*, 3 min. 20 sec., 8 mm Film-loop, Ealing Scientific, Watford WD2 4PW.
68. *Vibrations of a Drum*, 8 mm Film-loop, Ealing Scientific, Watford WD2 4PW.
69. *Vibrations of a Metal Plate*, 8 mm Film-loop, Ealing Scientific, Watford WD2 4PW.
70. *Problem Solving with Soap Films I*, 60 min., a video-tape of a lecture given by Dr. C. Isenberg, University of Kent at Canterbury.
71. *Problem Solving with Soap Films II*, 40 min., a video-tape of the different surfaces and bubbles produced by soap films, University of Kent at Canterbury.

VII *Educational Kits for Soap Film Demonstrations*

72. *Kubic Bubbles*, manufactured by Cochranes of Oxford Ltd., Leafield, Oxford, OX8 5NT.

VIII *Paintings*

73. Murillo, B. E. (1618–1682) *Boy Blowing Bubbles*. Two paintings mentioned in *Velasquez and Murillo*, C. B. Curtiss, Bouton, New York (1883), p. 280.
74. Chardin, J-.B. S. *Soap Bubbles*, oil (1739) National Gallery of Art, Washington D.C., also Metropolitan Museum of Art, New York, and William Nelson Gallery of Art, Kansas City.
75. Hamilton, W. *Bubbles*, water colour drawing, *circa* 1790, Arthur Gibey Collection, see *The Connoisseur*, **72**, 237 (Aug. 1925).
76. Manet, E. *Soap Bubbles*, oil (1867), private, see *Edouard Manet*, A. Jedleka, Rentsch, Zurich (1941), facing p. 87.
77. Millais, J. *Bubbles*, oil (1886), A. & F. Pears Ltd., London, see Roberts, Keith "John Everett Millais," *The Masters*, issue no. 67, Plate XVI.

IX   Specific references

78. Almgren, F. J. Jr. and Taylor, J. E. (July 1976) The Geometry of Soap Films and Soap Bubbles, *Scientific American*, 235, 1, pp. 82–93.
79. Bergmans, C. (1903) Joseph Plateau, *Biographie nationale*, 17, 768–788.
80. Bergmann, L. (1956) Experiments with Vibrating Soap Membranes, *Journal of the Acoustical Society of America*, 28, 6, 1043–1047.
81. Bragg, W. L. and Nye, J. F. (1947) A Dynamical Model of a Crystal Structure, *Proc. Roy. Soc. (London)*, 190A, 474.
82. Cockayne, E. J. (1970) On the Efficiency of the Algorithm for Steiner Minimal Trees, *SIAM J. of App. Maths.*, 18, 150.
83. Cockayne, E. J. and Melzak, Z. A. (1968) Steiner's Problem for Set Terminals, *Quart. Applied Maths.*, 26, 213.
84. Cook, G. A. (1938) Tough Soap Films and Bubbles, *Journal of Chemical Education*, 15, 161–166.
85. Delsaulx, J. (1884) Les travaux scientifiques de Joseph Plateau, *Revue des questions scientifiques*, 15, 114–158 and 16, 383–437.
86. Douglas, J. (1931) Solution of the Problem of Plateau, *Trans. American Math. Soc.*, 33, 263–321.
87. Dreyfus, S. E. and Wagner, R. A. (1972) The Steiner Problem in Graphs, *Networks*, 1, 195.
88. Einstein, A. (1901) Folgerungen aus den Capillaritätserscheinungen, *Annalen der Physik*, 4, 4, 513.
89. Fisher, L. R. and Oakenfull, D. G. (1977) Micelles in Aqueous Solution, *Chem. Soc. Reviews*, 6, 1, 20.
90. Gilbert, E. N. and Pollak, H. O. (1968) Steiner Minimal Trees, *SIAM J. App. Maths.*, 16, 1, 1.
91. Griffith, A. A. and Taylor, C. I. The Use of Soap Films in Solving Torsion Problems, *Superint. Roy. Aircr. Fact., Rep. Mem.* (n.s.), 333 (1917); *Tech. Rep. Adv. Comm. Aeron. (Brit.)*, 3 (1917–1918).
92. Goldschmidt, B. (1831) *Determinatio Superficei minimae rotatione curvae data duo puncta jungentis circa datum axem ortae*, Göttingen.
93. Hakimi, S. L. (1971) Steiner's Problem in Graphs and its Implications, *Networks*, 1, 113.
94. Hooke, R. (1672) On Holes (Black Film) in Soap Bubbles, *Communications to the Royal Society*, March 28. Also see: *History of the Royal Society*, Birch, T. (A. Millard, London), Vol. III, p. 29 (1757).
95. Isenberg, C. (1975) Problem Solving with Soap Films, *Physics Education*, Part I, 10, 6, 452–456; Part II, 10, 7, 500–503.
96. Isenberg, C. (1976) Problem Solving with Soap Films, *School Science Review*, 58, 202, 102.
97. Isenberg, C. (1976) The Soap Film: An Analogue Computer, *American Scientist*, 64, 5, 514.
98. Isenberg, C. (1976) Soap Films: The Analogue Solution to Some Practical Problems, *Proc. Roy. Inst. Gt. Brit.*, 49, 53–75.
99. Isenberg, C. (1977) Problem Solving with Soap Films, *Physics Teacher*, 15, 1, 9–18.
100. Kuehner, A. L. (1958) Long Lived Soap Bubbles, *Journal of Chemical Education*, 35, 337–338.
101. Lagrange, J. L. (1760–61) Essai d'une nouvelle méthode pour déterminer les maxima et les minima des formules intégrales indéfines, *Miscellanea Taurinensia*, 2, 173-195; *Oeuvres de Lagrange Bd. 1*, Gauthier–Villars, Paris, pp. 335–362 (1867).
102. Maxwell, J. C. and Rayleigh, Lord, Capillary Action, *Encyclopaedia Britannica*, 13th ed., Vol. 5, pp. 256–275.
103. Melzak, Z. A. (1961) On the Problem of Steiner, *Canadian Mathematical Bulletin*, 4, 143–148.

104. Mensbrugghe, G. Van der (1913) Joseph Plateau, *Liber Memorialis, Notices Bio-graphiques*, Tome II, Université de Gand, Faculté des Sciences, 54–71.
105. Mensbrugghe, G. Van der (1885) Joseph Antoine Ferdinand Plateau, *Nachruf. Annuaire Acad. Roy. Sci. Bruxelles*, **51**, 389–473.
106. Meusnier, J. B. M. C. (1785) Mémoire sur la courbure des surfaces, *Mém. Mathém. Phys. Acad. Sci. Paris*, prés. par div. Savans **10**, 477–510 (verlesen 1776).
107. Nicholson, M. (1949) The Interaction between Floating Particles, *Proc. Camb. Phil. Soc.*, **45**, 288.
108. Nitsche, J. C. C. (1974) Plateau's Problems and their Modern Ramifications, *American Math. Monthly*, **81**, 9, 945–968.
109. Overbergh, C. V. (1907–1908) *Le Movement scientifique en Belgique 1830–1905*, I, 368–396, Brussels.
110. Porter, A. W. (1964) Surface Tension, *Encyclopaedia Britannica*, **21**, 595.
111. Rayleigh, Third Lord (November 1944) Charles Vernon Boys 1855–1944, *Obituary Notices of Fellows of The Royal Society*, **4**, 771.
112. Reinold, A. W. and Rücker, A. W. (1883) The Limiting Thickness of Liquid Films, *Phil. Trans. Roy. Soc. (London)*, **174**, 645.
113. Reinold, A. W. and Rücker, A. W. (1877) On the Thickness of Soap Films, *Proc. Roy. Soc. (London)*, **26**, 334.
114. Satterly, J. (1947) Transverse Waves in Plane Soap Films, *Trans. Roy. Soc. Canada*, **41**, 55–61.
115. Schwarz, H. W. (1964) Rearrangements in Polyhedric Foam, *Recueil*, **84**, 771–781.
116. Skogen, N. (1956) Inverted Soap Bubbles—A Surface Phenomenon, *American Journal of Physics*, **24**, 239.
117. Trayer, G. W. and March, H. W. The Torsion of Members Having Sections Common in Aircraft Construction, *U.S. Nat. Adv. Comm. Aeron., Rept. 334 (1930)*; *Ann. Rep.*, **15**, 675 (1928–1929).
118. Velde, A. J. J. Van de (1948) Joseph Plateau 1801–1883, Briefwisseling met Adolphe Quetelet, chronologie en genealogie, *Mededelingen van de Koninklijke Vlaamse Academie voor Wetenschappen*, Letteren en schone Kunsten van België Klasse der Wetenschappen, jaargang X, No. 8, pp. 5–56.
119. Young, T. (1805) *Proc. Roy. Soc. (London)*, **95**, 65.
120. Young, T. (1805) *Phil. Trans. Roy. Soc. (London)*, **1**, 65, 84.
121. Strong, C. L. (1974, April) Scientific American **230**, 4, 116